The Getty Conservation Institute

T0133257

Alkoxysilanes and the Consolidation of Stone

George Wheeler

With Annotated Bibliography by
Elizabeth Stevenson Goins

2005

research in conservation

© 2005 J. Paul Getty Trust
Getty Publications
1200 Getty Center Drive, Suite 500
Los Angeles, California 90049-1682
www.getty.edu

Christopher Hudson, *Publisher*
Mark Greenberg, *Editor in Chief*

Patrick Pardo, *Project Editor*
Sheila U. Berg, *Copy Editor*
Pamela Heath, *Production Coordinator*
Hespenheide Design, *Designer*

Printed in Canada by Friesens

Reader's note: Color versions of figures 3.21, 3.22, 3.23, 3.24, and 5.4 can be accessed via the following URL: www.getty.edu/conservation/publications/pdf_publications.

Library of Congress Cataloging-in-Publication Data
Wheeler, George.
  Alkoxysilanes and the consolidation of stone / George Wheeler.
      p. cm. — (Research in conservation)
  Includes bibliographical references and index.
  ISBN-13: 978-0-89236-815-0 (pbk.)
  ISBN-10: 0-89236-815-2 (pbk.)
  1. Alkoxysilanes—Research. 2. Silicon—Synthesis. 3. Stone—Conservation
and restoration--Research. I. Title. II. Research in conservation (Unnumbered)
  TP245.S5W48 2005
  620.1'32—dc22
                                                              2004024444

For Sir Ralph and Michele

Contents

Foreword

Stone is one of the oldest building materials, and its conservation ranks as one of the major challenges of the field. This volume, *Alkoxysilanes and the Consolidation of Stone*, in the Getty Conservation Institute's Research in Conservation series, deals with the use of alkoxysilanes in the consolidation of stone. Although the literature on this topic is extensive, this is the first book to cover the subject comprehensively.

Alkoxysilanes have, in fact, a long history in stone conservation. As early as 1861 A. W. von Hoffman suggested their use for the deteriorating limestone on the Houses of Parliament in London. Between then and today, alkoxysilane-based formulations have become the material of choice for the consolidation of stone outdoors. Their basic properties, such as low viscosity and the chemical and light stability of the gels that form from them, have driven that choice. However, a gap inevitably exists between the ideal properties of a conservation material and its performance in practice. This book attempts to bridge that gap by synthesizing the copious and far-flung literature that ranges from syntheses of alkoxysilanes in the nineteenth century and the development of stone consolidation formulations in the twentieth century to the extensive contributions from sol-gel science in the 1980s and 1990s and the many (at times contradictory) reports of performance from the sphere of stone conservation.

George Wheeler clears a path for both researchers and practitioners and walks them through the strengths and limitations of alkoxysilanes as stone consolidants, maps out practical guidelines for their use, and posts trail markers for future development. Appended to the book is an annotated bibliography that will have a continuing life as a supplement to the AATA Online. I am convinced that conservators, scientists, and preservation architects in the field of stone conservation will find *Alkoxysilanes and the Consolidation of Stone* a valuable addition to their libraries.

I am grateful to George Wheeler, Director of Conservation Education and Research at Columbia University and Research Scientist at the Metropolitan Museum of Art and one of the world's leading experts on the science of stone conservation, for undertaking both the research on alkoxysilanes and the publication of this important material.

Timothy P. Whalen, Director
The Getty Conservation Institute

Preface

In this book I discuss alkoxysilanes only as consolidants for stone, although they have been used in the conservation of other materials such as stained glass (Schmidt and Fuchs 1991) and adobe (Chiari 1987), and they are found in many water repellents. Although properties of *hydrophobic* stone consolidants are examined, alkoxysilanes as water repellents per se are not discussed.

A few words should also be said about the definition of a consolidant. Consolidants have been used as a treatment for several conditions. An examination of the literature reveals that they have been used for such specific forms or manifestations of deterioration as scaling, flaking, spalling, and granular disintegration and the general condition of mechanically weakened stone. Both granular disintegration and mechanically weakened stone often result from the loss of intergranular cement, and it is the mending of this condition that is frequently the goal of consolidation treatments. The success of treatment with alkoxysilane-based consolidants depends on a thorough understanding of existing conditions and the suit ability of the consolidant for treatment of those conditions. The properties of an ideal consolidant and the ways that consolidation is achieved with different materials are a more complex terrain. Significant contributions to this discussion have been made by Sasse, Honsinger, and Schwamborn (1993); Butlin, Yates, and Martin (1995); and Price (1996). There is no consensus among these and other writers, however, and I leave fuller exploration of this subject for other works.

The primary audience for this book is conservators, but architects and scientists both in and outside conservation should also find it useful. It can also serve as an introduction to the subject for students of art conservation and historic preservation. To make the text more readable, I have relegated to the notes points of view that are more speculative.

The work is organized as follows. Chapter 1 is an overview of the literature on the use of alkoxysilanes as stone consolidants, with forays into chemistry, artists' materials, and sol-gel science. This chapter and the annotated bibliography at the end of the book owe much to the foundation laid by Grissom and Weiss (1981) in their own annotated bibliography.[1] They teased out most of the important threads, some quite difficult to locate, from an already complex fabric of literature. A significant amount of work has been produced since then, both in the conservation and the sol-gel literatures, the highlights of which appear in this chapter.

Chapter 2 owes a large debt to Brinker and Scherer's monograph *Sol-Gel Science* (1990), that is, the science of solutions that gel. That work synthesized what was a large and growing body of literature at that time. It shows the central role that tetraethoxysilane (TEOS) plays in sol-gel science and thus the importance of this literature to stone consolidation. In chapter 2 I further distill this body of work to the essential elements for understanding alkoxysilane stone consolidants.

Basic to that understanding is nomenclature, to which I devote a significant amount of space. Throughout the text, I avoid the term "silanes" as a catchall for alkoxysilane-based stone consolidants. As the title of this book indicates, the term used here is "alkoxysilanes." Silane is both the compound, SiH_4, and the root name for a class of silicon compounds, not just alkoxysilanes. The term "ethyl silicate" is often encountered in the literature. In this book, I use that term to refer to partially polymerized tetraethoxysilane. For TEOS-based consolidants discussed in this book, "partial polymerization" means various proportions of monomers, dimers, and trimers, up to approximately octamers. The term "alkali silicates" also appears in the literature for waterborne systems such as so-called water glass. Although often suggested as stone consolidants, these colloidal silicates are neither derived from alkoxysilanes nor are themselves alkoxysilanes and, therefore, are not discussed.[2]

Chapter 2 goes beyond nomenclature to discuss the influence of pH and catalysts on gel time and properties of the gel itself. This discussion takes place independent of and without particular reference to stone. Stone reenters the discussion in chapter 3, where I negotiate an inconsistent body of literature with particular attention to carbonate versus silicate rocks. Chapter 4 brings us onto less controversial ground and discusses the components of commercial and some noncommercial alkoxysilane-based consolidants. A subject making only a brief appearance in this book is consolidant formulations involving organic resins dissolved in alkoxysilanes. This should not be viewed as a negative judgment. On the contrary, excellent work on the use and explanation of effects of these consolidants has been done by Bradley (1985); Hanna (1984); Larson (1980); Miller (1992); Nonfarmale (1976); Rossi-Manaresi (1976); Thickett, Lee, and Bradley (2000); and Wheeler, Fleming, and Ebersole (1991). It becomes apparent from their investigations that the work of consolidation is done not by the alkoxysilane but by the organic resins, and this is the reason the subject receives such short shrift here.

Chapter 5 discusses the ways in which alkoxysilanes are used in practice and some of the difficulties arising from their use indoors, outdoors, and in conjunction with other conservation activities and materials. Tricks of the trade from the literature, from my experience in the field, and especially from other practitioners' experiences are passed along in this chapter.

Chapter 6 offers guidelines on the amount of time treatments with alkoxysilanes will last on different substrates. Long-term observation of the performance of conservation treatments—not only consolidants—is all too infrequent in our profession. In that regard I am indebted to many people at English Heritage and the Stone Conservation section of the

Building Research Establishment in Watford, England. This chapter would be very lean without the survey of Brethane treatments, recently published by Martin et al. (2002), that I use as the foundation for my discussion. Contributions in this area also have come from the Bayerisches Landesamt für Denkmalpflege in Munich and the geology department at the University of Munich; the Cesare Gnudi Center formerly in Bologna; the International Centre for the Study of the Preservation and Restoration of Cultural Property (ICCROM); and the Istituto Centrale per il Restauro in Rome. Among important observations, they all highlight the difficulty of designing laboratory testing procedures to evaluate current and future performance of stone consolidants. Despite the efforts of Marisa Laurenzi Tabasso and Jose Delgado Rodrigues, there is little consensus in thought or practice on how to evaluate consolidants in the laboratory. How can we acquire sufficient quantities of representative stone materials for testing? Which test procedures should be used? How does the test procedure relate to field conditions and the presumed mechanism(s) of deterioration at a given site? These fundamental questions are touched on in chapter 7 but await a more detailed treatment.

I naively approached this book with the idea of resolving all problems related to the use of alkoxysilanes as stone consolidants. This has not been possible, and many ambiguities remain. The annotated bibliography and what insights appear in the text are offered to assist researchers and practitioners to clarify the picture for themselves and for the conservation community. I cannot emphasize too strongly how important practicing conservators and restorers are to this process. Practitioners live in the field, the front lines of conservation and progress in understanding current treatments; improving these treatments and developing new ones cannot be done without their input.

Many of the bibliographic references come from so-called gray literature, such as conference papers. These publications usually do not undergo rigorous peer review, and in stone conservation they comprise a significant part of the literature—well over thirty volumes. There is a wealth of information in these volumes, but they are also replete with errors, incompleteness, and repetition. While these deficiencies do not entirely negate the value of the publications, they seriously undermine it. There is no substitute for rigorous peer review, which reveals our errors and strengthens our thinking, and I urge my colleagues to consider more frequent use of established journals for publication.

I began my work on alkoxysilanes in 1980 while interning at the Building Research Establishment under the guidance of Clifford Price. A risk in spending so much time with one conservation material is that one can lose perspective with regard to its usefulness. In this book I have sought to illuminate the subject rather than advocate for alkoxysilanes as consolidants for stone. Like many conservation materials, what initially seems a miracle elixir ultimately finds a place or a niche in an array of treatment possibilities. The goal of this book is to provide conservators, scientists, and architects with a better understanding of alkoxysilanes, their strengths and limitations, and how best to optimize their strengths and minimize their weaknesses.

Notes

1. Grissom and Weiss's *Alkoxysilanes in the Conservation of Art and Architecture: 1861–1981* is the supplement to vol. 18, no. 1, of *Art and Archaeology Technical Abstracts (AATA)*, then published by the International Institute for Conservation. Seymour Lewin's *The Preservation of Natural Stone, 1839–1965: An Annotated Bibliography* (1966), the supplement to *AATA*, vol. 6, no. 1, is also an important contribution.

2. Leaving these materials aside should not be taken as a blanket condemnation. Aqueous consolidants may be useful when dealing with salt-laden stone (see Mangio and Lind 1997; Kozlowski 1992). In more general terms, evaporating water puts less stress on the earth's environment than do organic solvents common to alkoxysilane consolidants. On the negative side, at least for the stone itself, are the alkalis—usually sodium, potassium, or ammonium—required to stabilize the large silicate particles from immediate gelation. These alkalis become carbonated or sulfated in air or polluted atmospheres, and the resulting salts can be damaging to stone. A recent publication that discusses the problem is Weeks 1998. The ammonium ions more common in today's aqueous silicates are at least volatile and may form fewer salts than sodium and potassium. Good introductions to the subject of alkali or metal salt silicates can be found in Iler 1979, *Soluble Silicates* (1982), and Brinker and Scherer 1990.

Acknowledgments

Many people helped me to get this book into its present form. First and foremost, I want to thank those who read the manuscript. Michele Marincola read the entire manuscript in several versions; Elizabeth Hendrix read an early version and offered insightful and detailed comments on both form and content; and Giacomo Chiari read later versions and gave me advice from the scientific perspective. Other colleagues read parts of the manuscript that fell into their area of expertise. Carol Grissom read the preface and chapter 1 and gave me detailed advice and corrections; George Scherer read several chapters and pointed out weaknesses in early versions of chapter 3; Rolf Snethlage and Eberhard Wendler improved chapters 3 and 4; and Clifford Price and Seamus Hanna made suggestions on chapters 5 and 6.

My scientific colleagues at the Metropolitan Museum of Art have been a resource for this book. We have written articles together, and some of that work found its way into these pages. They also provided some of the best images that appear in this book. I want to thank Api Charola, Bob Koestler, and Mark Wypyski, who were or are full-time members of the staff, and Juan Mendez Vivar and Elizabeth Goins, who were Fellows at the museum. Elizabeth also carried out masterful work that brought the annotated bibliography to fruition. I also want to thank Petria Noble, Susanne Ebersole, Li-Hsin Chang, and John Campbell, volunteers and students at the museum who produced valuable data. My Metropolitan Museum of Art colleague Rudy Colban obtained software that was very helpful in producing the manuscript.

The German literature has been vital to this book. I have received copious help from Michele Marincola, Marika Strohschneider, Mecka Baumeister, and Birte Graue in reading and translating this literature.

Over the years I have had discussions with colleagues and friends that have helped me to formulate many of the ideas presented here. In this regard I am particularly grateful to Rolf Snethlage, Eberhard Wendler, Ludwig Sattler, Stefan Simon, Giacomo Chiari, Anne Oliver, Frank Matero, David Boyer, Fran Gale, Norman Weiss, John Larson, Seamus Hanna, Sue Bradley, John Fidler, Tim Yates, Roy Butlin, Bill Ginell, Charles Selwitz, and Clifford Price.

I would once again like to thank Jeanne Marie Teutonico and John Fidler and the English Heritage staff for facilitating my tour of the Brethane sites in 1996. This work was a fundamental resource.

From the world of sol-gel chemistry, I want to thank my colleagues and friends Jeff Brinker, George Scherer, Sandi Fleming, and Jim Fleming for the many fruitful collaborations and discussions. I also want to thank the late Don Ulrich of the Air Force Office of Scientific Research for bringing me into the sol-gel group.

Many people have helped to obtain articles used in my research. Special thanks go to Jocelyn Kimmel, my intern in 1999, and to Linda Seckelson of the Watson Library of the Metropolitan Museum of Art who processed hundreds of interlibrary loan requests.

My dissertation adviser, Seymour Lewin, was a perfect mentor. He was a great teacher in all aspects of chemistry and a dedicated and insightful experimentalist.

The Getty Conservation Institute has patiently waited for the delivery of my manuscript. I thank all those with whom I have worked over the past years for that patience. I would particularly like to thank Sheila Berg for her excellent copyediting. The book has been significantly improved by her hand.

I also want to thank the Samuel H. Kress Foundation for the financial support I received through the American Institute for Conservation Publication Fellowship.

Special thanks also go to Joel Wiessler and the Thursday Night Group for helping me to recognize the forks in the road.

And finally, I want to thank Michele for going through her second first book.

Chapter 1

Historical Overview

Little fanfare accompanied the synthesis of silicon tetrachloride by J. J. Berzelius in 1824 (Berzelius 1824). Unlike the stunning in vitro synthesis of urea by his pupil Friedrich Wöhler only four years later, which destroyed the hallowed division between inorganic and organic compounds,[1] Berzelius's preparation slipped quietly into the annals of chemistry.

Over one hundred years would pass before silicon tetrachloride found its place as the building block for a new chemistry.[2] The long gestation period was punctuated by syntheses of other silicon-based compounds, including three alkoxysilanes that ultimately became important to stone conservation: tetraethoxysilane (Ebelmen 1846), methyltriethoxysilane (Ladenberg 1874), and methyltrimethoxysilane.[3] As early as 1861 A. W. von Hoffman suggested "silicic ether," a form of tetraethoxysilane (TEOS) or ethyl silicate, as a consolidant for stone—in this case, on the deteriorating fabric of the Houses of Parliament in London (fig. 1.1).[4]

In the 1920s ethyl silicate resurfaced in the conservation world with the work of A. P. Laurie, who experimented with several formulations and secured at least four patents for stone consolidants.[5] By 1932 disagreement arose over its efficacy, and in his classic text, *The Weathering of Natural Building Stone*, R. J. Schaffer (1932:88) dismissed Laurie's treatments, indicating that "silicon ester appeared to have exerted no protective effect."[6]

Figure 1.1

The Houses of Parliament in London. As early as 1861 A. W. von Hoffman suggested the use of tetraethoxysilane for the consolidation of the deteriorating Clipsham limestone on the Houses of Parliament, which had been constructed in 1839. Tetraethoxysilane had been first synthesized only fifteen years earlier by Ebelmen. Photo courtesy the Institute of Fine Arts, New York University.

In the meantime, ethyl silicate crept into other areas of art technology and conservation. King and Threlfall (1927, 1928, 1931); King (1930, 1931); Heaton (1930); and Gardner and Sward (1932) developed or patented paints using ethyl silicate as the medium. According to King, this "silicon ester binder" could be used for decayed stone, and Graulich (1933a, 1933b) similarly claimed that his ethyl silicate material could be used both as a paint medium and as a stone preservative. Soon thereafter, King and Warnes (1939) combined biocides with "silicic acid ester" to be used for masonry preservation.[7]

Other applications are cited in the literature through the 1940s and into the 1950s. Emblem (1947, 1948a, 1948b) outlined the chemistry of "silicon ester" under various reaction conditions and discussed its use as a paint and as a concrete and stone preservative. He found that when excess material was applied, a "white flurry" resulted that was "harmless and [could] be swept off" (Emblem 1947:240). Liberti (1955) found that efflorescences from the consolidant "disfigured" the surface when applied to brick. Cogan and Setterstrom (1947, 1948) provided an excellent review of both the chemistry and the uses of ethyl silicate and mentioned sites in the United States where it was applied to stone monuments.

In 1956 Harold Plenderleith published the first comprehensive text in English on the conservation of objects, *The Conservation of Antiquities and Works of Art.* In a section devoted to the consolidation of stone, he states, "for sandstone and siliceous limestone of large dimensions which are kept indoors, a most successful strengthening agent is silicon ester." Plenderleith's work may have reflected standard and pervasive conservation practice, but the literature at that time was not overflowing with examples of the application of alkoxysilanes to stone.[8] Shore (1957) mentioned "successful" treatment of sandstone, mortar, and chalk with ethyl silicate, and Smith (1957) performed consolidation on 238 sandstone objects with what he referred to as colloidal ethyl silicate.[9] An unattributed publication (Anon. 1959) at the Building Research Establishment offered a less favorable review of alkoxysilanes that echoed Schaffer's earlier evaluation.[10] A few years later Schaffer (Bailey and Schaffer 1964:n.p.) fired perhaps his final salvo on the subject: "All organized trials known to have been made with it, including those of A. P. Laurie who patented its use, have been discouraging. On stone of reasonably good quality it has had no apparent effect, good or bad. On stone of relatively poor quality its use has been followed by scaling of the treated surfaces."

The first mention of *alkyl*alkoxysilanes for stone preservation appears to be a U.K. patent application of the British Thomson-Houston Company (1947), where it is referred to almost in passing.[11] In 1956 Wagner averred that silicic acid esters were being replaced by silicones in Germany, and General Electric (1959) patented water repellents based on methyl- and ethylalkoxysilanes. Stone consolidants consisting of mixtures of ethyl silicate and methyltrialkoxysilanes appear in the work of Blasej (1959). Sneyers and de Henau (1963, 1968) reported on the consolidation of damaged calcareous and dolomitic stone with mixed alkoxy- and alkylalkoxysilanes, and Lerner and Anderson (1967) used similar materials as water repellents.

The late 1960s were a watershed for the study and use of alkoxysilanes on stone. It marked the initial work of Seymour Lewin (1966) at New York University who applied for a patent on a treatment system for carbonate rocks based on barium hydroxide. The next year he experimented with a new formulation based on ethyl silicate for "siliceous rocks."[12] Across the Atlantic Ocean, the team of Kenneth Hempel and Anne Moncrieff at the Victoria and Albert Museum focused on methyltrimethoxysilane (MTMOS).[13] Their interest centered on the resolidification or consolidation of "sugaring"[14] marble, and both field and laboratory experiments showed promising results: "Another winter's weathering has only confirmed the phenomenal properties of the material. The original piece of deteriorated marble exhibited no signs of breakdown whereas the untreated control sample has deteriorated to a pile of dust" (Hempel and Moncrieff 1972:174).[15]

Lewin (1972) and Hempel and Moncrieff (1972) presented the results of their work at a conference in Bologna[16] the same year that Plenderleith (Plenderleith and Werner 1971) published the new edition of his book. At the same time that enthusiasm for alkoxysilanes ran high for Lewin and Hempel and Moncrieff, Plenderleith's waned; he demoted ethyl silicate from "a most successful strengthening agent" (1956:305) to "a possible strengthening agent" (1971:310). Plenderleith made the distinction between sandstone and *siliceous* limestone, for which good results were achieved, and *fine* limestone, for which the results were not as good. Like Lewin, he suggests that ethyl silicate may not perform well on certain kinds of limestones.

While Schaffer in 1932 and 1964 may have sounded the first notes out of tune with the general melody of enthusiasm for alkoxysilanes, the 1970s saw a cacophony of conflicting reports. An unattributed contribution to the *New Scientist* (Anon. 1972), taken from a Building Research Establishment report (Clarke and Ashurst 1972), indicated failures of alkoxysilane stone preservatives. Aguzzi et al. (1973), in their discussion of treatments carried out with Lewin's ethyl silicate formulation, found poor penetration and the creation of a hard surface that was, nonetheless, more fragile to the touch, and Riederer (1971) ranged from ambivalence to epiphany. His tepid recommendation of alkoxysilanes for "smaller objects . . . when no other process can help" (p. 132) became, just two years later, high praise for the superior results obtained with ethyl silicate on the Alte Pinakothek in Munich and the Cathedral of Konstanz (Riederer 1972b). In 1973 Bosch echoed that judgment and again, with even greater resonance, in a later patent (Bosch 1976). By the mid- to late 1970s, Weber (1976:382–83) endorsed Wacker H on "highly absorbent and porous stone" because of its "penetration power," but Sleater (1977) found poor penetration on unweathered Indiana limestone, and Hoke (1976) detected uneven distribution in marble. At the close of the decade, Berti (1979) claimed excellent results for treatment of salt-laden brick, concrete, and sandstone with either straight alkyltrimethoxysilane or a mixed tetraethoxysilane–methyltriethoxysilane system. In contrast, Arnold (1978) cautioned *against* the use of a methyltrimethoxysilane consolidant on stone with high levels of salts.

The 1970s also proved important for stone conservation in general. The "stone conferences" initiated by V. Romanovsky were held for the first time in 1972.[17] Only one short article on alkoxysilanes (Bosch 1973) can be found in the proceedings of the first conference, which was held in La Rochelle, France, despite increasing use of the material at that time. In addition to the work cited above, the German patent was filed for the tetraethoxysilane- and tetraethoxysilane-methyltriethoxysilane–based stone consolidants now called Wacker OH and Wacker H, respectively (Bosch 1976). Evaluations of the similar Tegovakon products were published (Luckat 1972; Riederer 1973; Worch 1973), and an early reference was made to the product developed at the Building Research Establishment (BRE) and later called Brethane (Price 1975). Arnold, Price, and Honeyborne would separately or together publish several other articles on this consolidant over the next five years (Arnold 1978; Arnold, Honeyborne, and Price 1976; Arnold and Price 1976).

Before Hempel and Moncrieff's work, alkoxysilanes were used almost exclusively on outdoor stone.[18] Their work was continued and extended by John Larson, Hempel's protégé at the Victoria and Albert Museum. Working in a museum setting offered several advantages not available to outdoor practitioners, in particular, the ability to create and maintain favorable environmental conditions before, during, and after treatment.[19] One of Larson's additions to the Hempel and Moncrieff legacy touched on an important theme: he learned through experience that it was difficult to cohere larger pieces of carbonate stone previously consolidated with MTMOS. Larson (1980) concluded that the incorporation of organic resins, soluble in the alkoxysilane, would promote adhesion.[20] His excellent work in the late 1970s and early 1980s on the conservation of stone was crowned by the publication of his presentation at the Louisville stone conference, "A Museum Approach to the Techniques of Stone Conservation" (Larson 1983).

Larson's techniques achieved a kind of regal status in the museum community. The British Museum carried out extensive consolidation of deteriorated Egyptian limestone sculpture under the direction of Seamus Hanna, who had studied and worked with Larson. Hanna and British Museum conservation scientist Susan Bradley published important evaluations of treatments involving MTMOS (Hanna 1984; Bradley and Hanna 1985). The Metropolitan Museum of Art conserved deteriorated archaeological stone using Larson's methods, and staff members issued several publications on the ever-problematic Abydos reliefs (Charola 1983; Wheeler 1984; Koestler 1985).[21]

At about the same time Larson's seminal article was published, stone conservation literature and published references to the use of alkoxysilane stone consolidants experienced dramatic growth. The growth can be tracked by examining the semiannual abstracting journal *Art and Archaeology Technical Abstracts*. AATA contained 20 citations on stone conservation for a six-month period in 1966.[22] By 1985 more than 140 abstracts were collected for a similar six-month period,[23] and between 1980 and 2000 there were more than 200 articles on the subject of alkoxysilanes and stone conservation alone. This literature indicates that most consolidation treatments concerned sandstone, limestone, and marble.[24] When alkoxysilanes were used, treatment involved for the most part catalyzed TEOS- and mixed TEOS-MTEOS–based systems such as Wacker or *Conservare* H and OH, Tegovakon V and T, and catalyzed MTMOS systems such as Brethane as well as uncatalyzed MTMOS systems. The uncatalyzed systems often contained dissolved organic resins. Ethyl silicate formulations with or without MTEOS were commonly applied to sandstone, originally under the name *sandstein verfestiger*s,[25] or sandstone strengthening agents, and MTMOS formulations saw wider use on limestone and marble.[26] Sometime in the early 1980s ethyl silicate products began to be used with increasing frequency on limestone and marble—a practice that is quite common today.

The influence of different minerals on the performance of alkoxysilane consolidants became the focus of study beginning in the late 1970s. A. P. Laurie was probably the first to comment on the nature of gels formed by ethyl silicate in contact with stones of different mineralogies: he noted soft, weak, incoherent gels on limestone and hard, glassy, coherent gels on sandstone (Laurie 1926). Snethlage and Klemm (1978); Domaslowski (1988); and Goins (1995) each offered scanning electron micrographs (SEM) supporting the presence of linkages between grains of quartz in sandstones and alkoxysilane-derived gels, a theme seconded by Klemm, Snethlage, and Graf (1977) and Weber (1977). For calcite in limestones Domaslowski and Goins showed only isolated deposits of silica gel—neither connected nor conforming to the surrounding mineral grains. Further complicating the argument, Grissom et al. (1999) showed, also by SEM, conformal structures of gel derived from *Conservare* OH that provided large increases in strength to an argillaceous, fossiliferous, lime plaster; and Charola et al. (1984) showed connective strands of gel derived from MTMOS on Indiana limestone. Kumar (1995) examined the reactions of MTMOS in contact with silicate and carbonate rocks by Fourier transform infrared spectroscopy and concluded that there is no bonding in the case of limestone. Goins (1995) found nearly an eightfold difference in the percent increase in modulus of rupture of sandstone treated with Wacker OH over a purely calcitic limestone treated with the same consolidant. She concluded, however, that it is not possible to determine if this difference is due to bond formation in the silicate rock. Wheeler and colleagues (Wheeler et al. 1991; Wheeler, Fleming, and Ebersole 1992) also looked at increases in modulus of rupture provided by MTMOS to a quartz-rich sandstone and a purely calcitic limestone. A consistent trend toward greater strength increases to sandstone was demonstrated, and the conclusion drawn from these data, if only by inference, was that these increases in sandstone are the result of adhesion between the consolidant and the minerals comprising the rock.

Wheeler et al. (1991) also found that mixtures of ACRYLOID B72 acrylic resin[27] and MTMOS exhibited a significant slowing of polymerization of MTMOS that promoted its evaporation. Bradley (1985) also demonstrated that evaporative loss during treatment of carbonate rocks with MTMOS was a problem. Charola et al. (1984) determined the amount and condition of the gel formed with neat MTMOS in the absence of stone. They showed that the amount deposited (or the amount lost due to evaporation) and the physical integrity of the gel were linked to the relative humidity to which the reacting alkoxysilane was exposed: low relative humidity gave low yield but uncracked, monolithic gels; high humidity gave higher yield but cracked or fragmentary gels. Wheeler et al. (1991) went on to show that for purely calcitic limestone, MTMOS functioned almost entirely as a solvent for organic resins rather than as an additional consolidant. The net result of the work of Bradley, Charola, Wheeler and their coworkers was that MTMOS and MTMOS–organic resin mixtures were used less frequently for the consolidation of carbonate rocks kept indoors and were replaced by more traditional resin-solvent mixtures such as ACRYLOID B72 acrylic resin in acetone, acetone-ethanol, acetone-diacetone alcohol, and toluene.[28]

It was realized early on that evaporative loss of MTMOS or other alkoxysilanes could be limited by decreasing the gel time through the addition of catalysts. Arnold, in outlining some of the properties of Brethane, provides an excellent description of how field experience led to the idea of reducing gel time:

> The systems [in this case the system was Rhone-Poulenc's X54-802, i.e., MTMOS] that were tried initially, took several days, or even weeks, to solidify, and during this time the preservative penetrated over-deeply into the stone. The amount of preservative in the critical outer 25 mm was so severely depleted that very little remained. [This depletion is due not only to "over-deep" penetration but also to evaporation.] It was clearly essential to decrease the time required for curing, but it was equally important that this should not be achieved at the expense of any increase in viscosity during the time taken to apply the preservative. A system has now been developed (British Patent Application 31448/76) which retains its initial low viscosity for several hours but which then gels very rapidly, preventing further dispersion. (1978:4–6)

Several other commercial formulations such as Wacker, *Conservare*, Keim, and Tegovakon also employ catalysts to reduce gel time.

By the 1980s it was possible to reevaluate the condition of earlier treatments—a regrettably infrequent occurrence in the field of conservation. Rossi-Manaresi's *Conservation Works in Bologna and Ferrara* (1986), a companion to the Preprints of the 1986 International Institute for Conservation conference, *Case Studies in the Conservation of Stone and Wall Paintings*, offered a guided tour of treatments to stone sculpture, monuments, and buildings in these Italian cities. She began such evaluations in *The Conservation of Stone II* (Rossi-Manaresi 1981) and continued this tradition through the rest of the 1980s and the 1990s. Regarding several sandstone reliefs treated with either Wacker H or the so-called Bologna Cocktail,[29] Rossi-Manaresi observed that the Wacker product had continued to provide a brittle solidity for flakes of stone on the surface, but the flakes themselves had not remained well adhered to the rest of the stone. In Germany, Sattler and Snethlage (1988) returned to St. Peter's Church in Fritzlar to evaluate the 1979 "silicon acid ester" (ethyl silicate) consolidation treatment. Their evaluation consisted in biaxial mechanical testing of slices from large cores removed from the building. They concluded that the stone strengthening achieved in the treatment of 1979 was still present in 1988.

Before 1980 sparing reference is made to the use of alkoxysilanes on stones other than sandstone, limestone, and marble, and treatments on the two sedimentary rocks far outweigh those on marble. We find Riederer (1974) recommending ethyl silicate for salt-ridden granite[30] in Egypt, and Plenderleith (1956:305) indicating "disappointing results [with ethyl silicate] in the case of lavas." Although the number of published studies on other stone types increased after 1980, they are still limited. Rossi-Manaresi and Chiari (1980) evaluated the treatment of a volcanic tuff from Ecuador with acid-catalyzed ethyl silicate. Nishiura (1987) attempted to consolidate crushed Oya stone, also a volcanic tuff,[31] with methyltriethoxysilane and with alkoxysilane coupling agents[32] and concluded, like Rossi-Manaresi, that the alkoxysilane by itself could not "fix" granules of the stone. Our knowledge and understanding of the conservation of volcanic rocks have been expanded in an excellent volume, *Lavas and Volcanic Rocks* (1994),[33] edited by Charola, Koestler, and Lombardi, the only book-length publication on these frequently deteriorated and difficult-to-preserve rock types. The impetus for this work was the study of the megalithic sculptures on Easter Island. The four articles in this volume that address consolidation treatments in some way concern alkoxysilanes. Useche (1994), for example, determined that both Wacker H and OH gave good strength increases to the volcanic tuff used for the portal of the Church of Santo Domingo in Colombia but also darkened the stone to some degree.[34] Lukaszewicz (1994) showed that these same products reduced open porosity in three central European tuffs, and Tabasso, Mecchi, and Santamaria (1994) found moderate improvement in the compressive strength of volcanic tuffs with Tegovakon V, another ethyl silicate–based consolidant. Wendler, Charola, and Fitzner (1996) studied the treatment of volcanic tuffs from Easter Island with ethyl silicates, including newly developed formulations with elastomeric components. Rodrigues and Costa's edited volume, *Conservation of Granitic Rocks* (1996), is a well-crafted work on the petrography, deterioration, and treatment of granites used for monuments in Spain and Portugal. Among the consolidants examined was Wacker OH, which gave good increases in ultrasonic velocity.[35] However, samples of these granites failed to retain the increases when exposed to wet-dry cycling.

Little mention of other stone types treated with alkoxysilanes is found in the conservation literature. Without question, consolidation has been performed on other stones, but the results have not been published.

Although the range of stones studied may seem narrow, it should be emphasized that the frequency of consolidation treatment of certain stone types has mostly to do with the frequency of their use and their tendency to deteriorate. These conditions—frequency of use and tendency to deteriorate—are pertinent for many sandstones and limestones which themselves exhibit great variation in composition and structure. Reports on the performance of alkoxysilanes on sandstone and limestone also vary widely: the large strength increases on quartz-rich sandstones versus the limited strength increase on purely calcitic limestones demonstrated by Wheeler et al. (1991, 1992) and Goins (1995); warnings by Felix (1995) and Felix and Furlan (1994) of excessive shrinkage in some clay-bearing sandstones; and, finally, the tantalizing comments by Butlin, Yates, and Martin (1995) and Martin et al. (2002) that field experience in England indicates that alkoxysilanes perform better on limestone than on sandstone.

It is appropriate that we come to the end of this historical overview on a note of uncertainty concerning the use of various alkoxysilane formulations on different stone types. I have barely differentiated among the many commercial and ad hoc formulations—what they contain and how those contents affect performance—and have only hinted at a relationship between the performance of these consolidants and a stone's mineralogy and structure. Each of these topics is addressed in later chapters. The basic chemistry and physics of alkoxysilanes and their gels must first be laid out.

Notes

1. See Bensaude-Vincent 1996; Partington 1964. Before Wöhler's synthesis it was believed that organic compounds could be produced only by living beings.

2. Silicon tetrachloride would become an important building block for the silicone resin industry that flourished in the United States and the Soviet Union. In the United States the development of silicones was fertilized and carefully tended by the chemist Eugene G. (George) Rochow at General Electric. See Müller 1965; Rochow 1995.

3. A complete reference cannot be found, but it was probably first synthesized by Frederic Stanley Kipping in about 1904.

4. Von Hoffman (1861) would refer to ethyl silicate or partially polymerized tetraethoxysilane. As a group, alkoxysilanes are referred to in the literature in a number of ways: silicon esters, silicic acid esters, ortho silicates, alkyl silicates, and, of course, alkoxysilanes. For tetraethoxysilane, the following names might be encountered: ethyl silicate, tetraethylorthosilicate, silicic acid ethyl ester, and, succumbing to the scientist's irrepressible compulsion to alphabetic abbreviation, TEOS.

5. Laurie 1923, 1925, 1926a, 1926b. Around the same time, de Ros and Barton (1926) patented the direct application of silicon tetrachloride for stone preservation. There appears to have been little follow-up to this work, possibly for two reasons: because of the difficulty of handling silicon tetrachloride due to its volatility and because reaction of silicon tetrachloride with water yields hydrochloric acid as a byproduct, with obvious complications for the applica-

tion to any stone containing acid-soluble materials, such as calcite and dolomite.

6. A 2004 reprint edition, with an introduction by Timothy Yates, is available from Donhead Publishers, and the information it contains continues to be valuable.

7. This same approach was tried in the 1980s when ProSoCo in Lawrence, Kansas, marketed a stone consolidation product called *Conservare* H40 Plus. Some researchers observed the development of a pronounced yellow color with this product. See Koestler and Santoro 1988; Tudor 1989.

8. The body of conservation literature was not vast in 1956, which made Plenderleith's book even more valuable as a handbook of practice. Even today his book remains the only serious attempt in English at a comprehensive text on conservation. Despite the thinness of the literature at this time, it is possible that alkoxysilane treatments were executed on stone but went unrecorded—a practice not without precedence in conservation.

9. It is not clear from Smith (1957) whether the product, manufactured by Shaw Processes Ltd., Newcastle-upon-Tyne, was truly collodial and, therefore, outside the purview of this book. Colloids are collections of particles that are less than a micron and more than a millimicron in diameter, that is, just below the resolution of a light microscope down to molecular dimensions. Concentrated colloidal silica solutions must be stabilized against coagulation, and in the case of silicates this often consists of ionic bases such as ammonium, potassium, and sodium hydroxides. These colloidal liquids are thus more like water glass with the attendant problems noted in the preface.

10. At this time Schaffer was employed by the Building Research Station (BRS), later called the Building Research Establishment (BRE). The BRE is located in the Watford area of Hertfordshire. At the BRE, significant work on the conservation of stone has been, and is still being, carried out.

11. Stone and Teplitz (1942) had earlier tested alkylsilicates for the consolidation of earth.

12. Implicit in Lewin's creation and specification of different treatments for rocks with different mineralogies is the fact that it may be difficult, if not impossible, to find a single treatment for all stone types.

13. They reported on the use of Rhone-Poulenc's X54-802, which is methyltri-methoxysilane.

14. "Sugaring" is a manifestation of stone deterioration in which grains of stone become loose due to dissolution along grain boundaries, loss of cementing material, or separation along grain boundaries induced by heat, freezing water, and salt crystallization. With many marbles, the grains are about the size (and, of course, are the same color) as grains of table sugar. This form of deterioration is a type of granular disintegration. For an excellent review and classification of forms of weathering and deterioration, see Fitzner, Heinrichs, and Kownatzki 1995.

15. Hempel and Moncrieff's paper opens with the comment, "Since reporting to the Bologna Meeting two years ago we have made only little progress in the matter of marble consolidation." This indicates that much of their work had been completed by 1969. No proceedings of the 1969 Bologna meeting appear to be available.

16. The publication was the first of five conference proceedings edited by Rossi-Manaresi as part of her major contribution to the stone conservation literature.

17. The first of these conferences was in La Rochelle (1972), with Romanovsky as host. It has been followed by Athens (Th. Skoulikidis) in 1976, Venice (M. Marchesini) in 1979, Louisville (K. Lal Gauri) in 1982, Lausanne (Vincio Furlan) in 1985, Torun (W. Domaslowski) in 1988, Portugal (Jose Delgado Rodrigues) in 1992, Berlin (Josef Riederer) in 1996, Venice (Vasco Fassina) in 2000, and Stockholm (Marie Klingspor Rotstein) in 2004.

18. The two exceptions are the enigmatic reference by Plenderleith that silicon ester should be used *only* on stone kept indoors; and Smith 1957.

19. Larson did not, however, confine his use of MTMOS-acrylic mixtures to indoor settings.

20. Larson employed three resins: Rohm and Haas ACRYLOID B44 (methyl-methacrylate), Rohm and Haas ACRYLOID B72 (ethylmethacrylate-methylacrylate), and Racanello E55050, the so-called acrylic-silane (a mixture of methylphenylsilicone and ACRYLOID B67—*i*-butylmethacrylate—in toluene [analysis provided by Susan Bradley of the British Museum]). This subject of adhesion was taken up again by Dinsmore (1987); the use of mixtures of resins and alkoxysilanes had been explored earlier by Munnikendam (1972).

21   The long history of conservation treatment of these Eighteenth Dynasty lime-stone reliefs began in 1912. Deterioration by salt crystallization continued through most of the twentieth century. The most recent campaign of treatment began in 1981 and ended in 1992 with the installation of most of the reliefs in the Temple of Dendur wing of the Metropolitan Museum of Art. Several conservators worked on the reliefs during this period, including Jennifer Dinsmore, Constance Stromberg, Leslie Ransick, Susanne Ebersole, Sara MacGregor, and Sarah Nunberg. Nunberg (1996) continued to study the problem of salt decay and its amelioration on the Abydos reliefs.

22. In the same year, however, Lewin (1966a) published *The Preservation of Natural Stone*, a bibliography containing more than 300 citations, one of several annotated bibliographies on art materials that he published in the 1960s when he was a professor at New York University's Conservation Center.

23. The conservation literature grew significantly during this time. A contributing factor, at least in the United States, was the birth of the art conservation training programs attached to academic institutions. The program at New York University began in 1960, at the State University of New York in Cooperstown in 1970, and at the University of Delaware (Winterthur) in 1974. A short-lived program began at Oberlin College in 1969. The Fogg Art Museum started formal training in 1927 that continues today only in the form of an advanced internship program. See *North American Graduate Programs in the Conservation of Cultural Property* (2000).

24. Sandstone, limestone, and marble have been used frequently in sculpture, monuments, and buildings because they are abundant at the earth's surface and are relatively easily extracted and carved. Sandstone and limestone are mechanically weaker than many other rock types, and this weakness, combined with their higher porosity, makes them susceptible to deterioration by salt crystallization, freezing water, and, in some cases, wetting and drying. Limestones, marbles, and calcareous sandstones are also susceptible to deterioration by acid rain and dry deposition. The tendency for all these rocks to deteriorate makes them likelier candidates for consolidation treatment.

25. Although this term appeared on early labels, Snethlage (pers. com.) suggests that the term is too forceful, perhaps even indicating overconsolidation (discussed in later chapters).

26. The sandstone/limestone division is to some degree geographic. The TEOS formulations were developed largely in Germany, where sandstone is not only dominant but also often badly deteriorated. The MTMOS formulations were developed in England. England is, of course, not without sandstone, but limestone is more commonly used for outdoor monuments there than in Germany. In the British Museum limestone objects were treated with MTMOS formulations and sandstone objects were treated with TEOS formulations such as Wacker OH.

27. ACRYLOID B72 is also referred to as PARALOID B72.

28. Lambertus Van Zelst commented that MTMOS is nonetheless a very good solvent: it dissolves many acrylic and polyvinyl acetate resins used in conservation and has low viscosity and relatively low vapor pressure (see chap. 2).

29. Bologna Cocktail is the informal name of a consolidant formulation developed by Nonfarmale and Rossi-Manaresi. It contains both acrylic and silicone resins dissolved in a mixture of solvents. See Nonfarmale 1976; Rossi-Manaresi 1976.

30. The mineralogy of many granites is similar to that of many sandstones. In fact, one group of sandstones is referred to in the geological literature as arkose, which, loosely translated, means "granite wash," indicating the detrital material that formed the sediment that eventually became the sandstone was from the weathering of granite.

31. Richard Newman characterized Oya stone as ignimbrite or tuff-lava (Wheeler and Newman 1994). Because of its poor quality, Nishiura would have little difficulty crushing this stone. Frank Lloyd Wright specified Oya stone for the Imperial Hotel constructed between 1916 and 1922 in Tokyo. The Imperial Hotel survived the devastating earthquake that hit Tokyo in 1923 as well as the American saturation bombing campaigns of 1945. It was the slower processes of destruction of the stone—environmental deterioration due to the stone's high porosity and mechanical weakness—that ultimately led to the demolition of the hotel in 1967.

32. A coupling agent is an adhesion promoter employed in composite materials. It has two functional groups with disparate bonding affinities: one has an affinity for organic resins; the other has an affinity for an inorganic filler such as glass. Many alkoxysilanes are also coupling agents, a good example being aminopropyltrimethoxysilane, $NH_2CH_2CH_2CH_2Si(OCH_3)_3$. The amine group would be attracted to an organic resin such as epoxy or polyester, while the methoxysilane would bond to a filler such as glass or quartz.

33. Sculptures, monuments, and buildings constructed with these rock types are found in areas of current or historic volcanic activity. These include all areas along the Pacific Rim extending from the southwest coast of South America to the northwest coast of North America, extending west to Japan, and south to Indonesia; the central Mediterranean from Sicily up to the Italian mainland near, of course, Pompeii; and regions of central Europe such as Germany and Poland.

34. Wheeler and Newman (1994) observed darkening in Oya stone that took several weeks to disappear.

35. There is often a good correlation between increases in ultrasonic velocity and increases in mechanical properties such as tensile and compressive strength, as well as modulus of rupture. There is a direct correlation between ultrasonic velocity and elastic modulus.

Chapter 2

The Chemistry and Physics of Alkoxysilanes and Their Gels

What accounts for the nearly century-and-a-half interest in alkoxysilanes for the consolidation of stone? Two properties are often cited: its low viscosity and its ability to form siloxane (Si-O-Si) bonds. Low viscosities make for mobile liquids that easily invade stone's intergranular network (table 2.1; fig. 2.1). Siloxane bonds are relatively strong, possess thermal and oxidative stability, and resist cleavage by ultraviolet solar radiation (Smith 1983:16); this resistance is confirmed by the abundance and long life of rain- and sun-drenched silicate minerals in the earth's crust (see table 2.2).[1] Like these silicate minerals, the gels formed from the alkoxysilanes shown in table 2.1 have little tendency to discolor through breakdown and reconfiguration of the bond network. The inherent light stability, which contrasts sharply with many organic resins, makes siloxane systems attractive for use outdoors.

Low viscosity of the starting materials and stability of the end products may be desirable properties for stone consolidants, but how are alkoxysilanes transformed into consolidating gels, and what properties or conditions mediate that transformation? I will begin to answer these questions by examining in isolation the essential ingredients, the alkoxysilanes.

Table 2.1

Viscosities in Pa·sec x 1000 at 25°C

Alkoxysilanes	Viscosities
Methyltrimethoxysilane	0.3750
Ethyltrimethoxysilane	0.4991
i-butyltrimethoxysilane	0.8078
1,2 dimethyltetramethoxydisiloxane	0.9554
Methyltriethoxysilane	0.5829
Tetraethoxysilane	0.7180
Ethanol	1.0826
Water	0.8903

Source: The data were generated in contract research supported by the Getty Conservation Institute.

Table 2.2

Bond strengths in kcal/mol

Si—O	108
C—H	99
C—C	83
C—O	86
C=C	148
C=O	169

Source: Eaborn 1960.

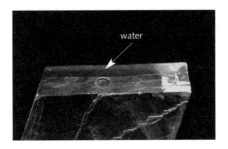

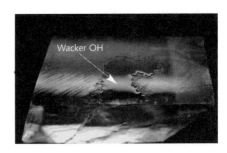

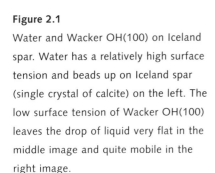

Figure 2.1

Water and Wacker OH(100) on Iceland spar. Water has a relatively high surface tension and beads up on Iceland spar (single crystal of calcite) on the left. The low surface tension of Wacker OH(100) leaves the drop of liquid very flat in the middle image and quite mobile in the right image.

Essential Ingredients

The nomenclature for silicon compounds such as alkoxysilanes is addressed first.[2] The root name is *silane*, SiH_4, derived by analogy with methane, CH_4. Working from this root, each hydrogen may be replaced by other elements or groups. For example, replacement by one, two, three, or four chlorine atoms—$ClSiH_3$, Cl_2SiH_2, Cl_3SiH, $SiCl_4$—yields the names chlorosilane, dichlorosilane, trichlorosilane, and tetrachlorosilane, the latter more often called silicon tetrachloride. Implicit in this naming system is the fact that silicon has four pendant groups, and, unless specifically named, these pendant groups are assumed to be hydrogen. Thus the compound CH_3SiH_3 is called methylsilane, that is, one named methyl group and three unspecified hydrogens.

Pendant elements or groups may be reactive, and the number of reactive groups determines a compound's functionality. For our purposes, this reactivity or functionality refers to their propensity for hydrolysis—in effect, for being removed by water and replaced by a hydroxyl (OH) group:[3]

Hydrolysis   Si-X + H_2O => Si-OH + HX

One of the products of this hydrolysis reaction is Si-OH, or silanol (like alcohol). Silanols can react with one another to produce siloxane bonds in a condensation reaction:[4]

Condensation   Si-OH + HO-Si => Si-O-Si + H_2O

The most common reactive groups are H, F, Cl, and RO, where R is the general symbol for alkyl groups such as methyl, CH_3, or ethyl, CH_3CH_2. These RO groups are called alkoxy—*alk*yl + *oxy*gen—for example, methoxy, CH_3O, and ethoxy, CH_3CH_2O. Following the nomenclature system described above, the compounds $(CH_3O)_3SiH$ and $(CH_3CH_2O)_3SiH$ are called trimethoxysilane and triethoxysilane, respectively. Replacing all four hydrogens with alkoxy groups produces such compounds as tetramethoxysilane, $Si(OCH_3)_4$, usually abbreviated TMOS, and tetraethoxysilane, $Si(OCH_2CH_3)_4$, or TEOS.

Pendant groups may also be unreactive. Organic groups directly attached to silicon by means of a Si-C bond, that is, not interposed by oxygen, remain unreactive or stable to hydrolysis. In the compound methylsilane, CH_3SiH_3, or methyltrimethoxysilane, $CH_3Si(OCH_3)_3$, the hydrogens or methoxy groups will react with water, while the methyl groups directly attached to silicon remain so throughout hydrolysis and condensation:

Hydrolysis
$$CH_3Si(OCH_3)_3 + 3H_2O \Rightarrow CH_3Si(OH)_3 + 3CH_3OH$$

Condensation
$$CH_3Si(OH)_3 + (HO)_3SiCH_3 \Rightarrow CH_3\text{-}\overset{OH}{\underset{OH}{Si}}\text{-}O\text{-}\overset{OH}{\underset{OH}{Si}}\text{-}CH_3 + H_2O$$

By mixing and matching reactive and unreactive groups, hundreds of compounds can be created that form siloxane bonds.[5]

Surprisingly, only a limited number of such compounds have been used with any regularity for the consolidation of stone: methyltrimethoxysilane (MTMOS), methyltriethoxysilane (MTEOS), and tetraethoxysilane (TEOS).[6] To function as a consolidant, compounds must have the ability to form a three-dimensional network and, therefore, must have a minimum of three reactive groups. This requirement eliminates all difunctional compounds that form only linear polymers.[7] In addition, some tri- and tetrafunctional compounds have basic properties that disqualify them as either stone consolidants or conservation materials. Actual silanes such as SiH_4 and $R\text{-}SiH_3$ (also referred to as silicon hydrides) are toxic, volatile,[8] and generate hydrogen gas on hydrolysis:

$$SiH_4 + 4H_2O \Rightarrow Si(OH)_4 + 4H_2$$

Tri- and tetrafunctional chloro- and fluorosilanes (or silicon halides) are also volatile,[9] and generate hydrochloric and hydrofluoric acids on hydrolysis—the former damaging to carbonate rocks such as limestone, travertine, and marble; the latter, to virtually all stone:

$$SiCl_4 + 4H_2O \Rightarrow Si(OH)_4 + 4HCl$$

$$SiF_4 + 4H_2O \Rightarrow Si(OH)_4 + 4HF$$

On the other hand, tri- and tetrafunctional *alkoxy*silanes are generally lower in toxicity and volatility or vapor pressure (see table 2.3), and the byproducts of their hydrolysis are alcohols, which are not corrosive to stone:

$$Si(OR)_4 + 4H_2O \Rightarrow Si(OH)_4 + 4ROH$$

Compared to silicon hydrides and silicon halides, tri- and tetraalkoxysilanes generally have a low to moderate reactivity to water, which can be an advantage for a stone consolidant. Too rapid a reaction may limit depth of penetration of the liquid into the stone before gelation occurs. In fact, the list of gel-forming silicon compounds possessing the requisite reactivity comprises only a few compounds: tetramethoxysilane, methyltrimethoxysilane, methyltriethoxysilane, tetraethoxysilane.[10] However, qualifying a compound for stone consolidation under the rubric "reactivity" disguises the delicate balance of properties embodied in some alkyltrialkoxysilanes, $R'\text{-}Si(OR)_3$, and tetraalkoxysilanes, $Si(OR)_4$, a balance that arises from the selection of the different alkyl and alkoxy groups.

Table 2.3

Vapor pressures in mm Hg at 25°C (molecular weights in amu are in parentheses). To obtain values in Pascals multiply by 133.32.

Methyltrimethoxysilane	(136.22)	31
1,2-dimethyltetramethoxydisiloxane	(226.38)	7
Tetramethoxysilane	(152.22)	15
Ethyltrimethoxysilane	(150.25)	14
n-propyltrimethoxysilane	(164.27)	12
i-butyltrimethoxysilane	(178.30)	6
Methyltriethoxysilane	(178.30)	8
n-octyltriethoxysilane	(276.48)	2
Tetraethoxysilane	(208.33)	5
Silicon tetrachloride	(169.90)	194 (20°C)
Acetone	(58.08)	231
Ethanol	(46.07)	59
Toluene	(92.14)	29

Source: Data produced by the author in contract research for the Getty Conservation Institute.

We may begin to understand why the list of qualified compounds is so short by examining the influence of R' and OR on some basic properties. The smallest possible groups are methyl and methoxy, CH_3 and CH_3O, leading to the compounds methyltrimethoxysilane, $CH_3Si(OCH_3)_3$, or MTMOS, and tetramethoxysilane, $Si(OCH_3)_4$, or TMOS. One important property for a consolidant is volatility or vapor pressure. Too high a vapor pressure leads to excessive evaporation of the consolidating material. Table 2.3 lists the vapor pressures of several alkoxysilanes along with some common solvents used in conservation.

As might be expected for the compound with the *lowest* molecular weight, MTMOS has the *highest* vapor pressure of all the alkoxysilanes listed, 31 mm Hg. This vapor pressure is close to that of toluene —29 mm Hg—ironically, a solvent often used in conservation because of its slow evaporation.[11] Unlike toluene, MTMOS hydrolyzes and condenses, and both scientists and conservators recognized early on that it produced a gel simply in the presence of water vapor (Charola, Wheeler, and Freund 1984; Hempel and Moncrieff 1972; Larson 1980). Under the right conditions MTMOS reacts quickly enough to overcome its tendency to evaporate. Both the higher vapor pressure and the rate of reaction of MTMOS depend in part on the smaller size and mass of the methyl and methoxy groups.

Brinker and Scherer, in their seminal book *Sol-Gel Science* (1990) cite two other properties that come into play in the balancing act of alkoxysilane reactivity: steric effects and inductive effects (see also Voronkov, Mileshkevich, and Yuzhelevski 1978). Steric effects, or steric hindrance as it is sometimes called, consist in the crowding or blocking of the central silicon atom by larger or geometrically more complicated alkyl or alkoxy groups. This crowding reduces the rates of hydrolysis and condensation (Aelion, Loebel, and Eirich 1950). Without hydrolysis there can be no condensation, and without condensation, no gelation. Following this logic, methyltri*methoxy*silane hydrolyzes more rapidly than does methyltri*ethoxy*silane, which, in turn, reacts more rapidly than methyltri-*n-propoxy*silane. The same logic applies to tetraalkoxysilanes, where, at least under neutral conditions, the relative hydrolysis rates are tetramethoxysilane > tetraethoxysilane > tetra-*n*-propoxysilane (Schmidt, Scholze, and Kaiser 1984).

Inductive effects result from substituting *alkyl* for *alkoxy* groups, for example, replacing a methoxy group on tetramethoxysilane with a methyl group, thereby transforming it into *methyl*trimethoxysilane. These substitutions increase the electron density on the silicon atom and, under acid conditions, increase the rate of hydrolysis. For base conditions, the alkyl-for-alkoxy trade decreases the rate of hydrolysis (Brinker and Scherer 1990:122–23; Schmidt, Scholze, and Kaiser 1984:1–11).

Some sense of the combined influence of vapor pressure, steric hindrance, and inductive effects can be gained by exposing several alkoxysilanes to the same, simple reaction conditions: atmospheric moisture in the laboratory. Table 2.4 indicates the weight/weight percent mass and qualitative condition of the given alkoxysilane after a reaction time of 120 days.

Table 2.4

Percent mass return[a] of neat alkoxysilanes reacted at laboratory conditions, approx. 20°C and 40% RH, for 120 days

Methyltrimethoxysilane (MTMOS)	13%	brittle solid
Ethyltrimethoxysilane (ETMOS)	40%	rubbery solid
n-propyltrimethoxsilane (*n*-PTMOS)	0%	
i-butyltrimethoxysilane (*i*-BTMOS)	20%	liquid[b]
Tetramethoxysilane (TMOS)	<1%	brittle solid
Tetraethoxysilane (TEOS)	0%	
n-octyltriethoxysilane (*n*-OTEOS)	98%	liquid

Source: Data were generated in contract research for the Getty Conservation Institute.

Notes:
[a]Percent mass return is defined as the mass of gel that forms divided by the mass of the original liquid times 100%. We can calculate a theoretical percent mass return for a given compound assuming complete hydrolysis and condensation and no evaporation of the alkoxysilane. For example, for methyltrimethoxysilane, complete hydrolysis and condensation are summarized by the following reaction:

$$CH_3Si(OCH_3)_3 \text{ (136 amu)} => CH_3Si(O)_{1.5} \text{ (67 amu)} + 3CH_3OH$$

If we assume that the methanol evaporates, then the percent mass return is $(67/136) \times 100\% = 49\%$. Incomplete hydrolysis may lead to percent mass returns *higher* than this number, but only evaporation of MTMOS monomers or oligomers can lead to returns *lower* than 49%. The numbers in this table cannot be considered absolute for each compound. They vary with temperature, relative humidity, and the surface area:volume ratio of the liquid.

[b]After several more months, all of this liquid evaporated.

Examining the series of alkyltrimethoxysilanes is instructive. These compounds differ successively by the addition on average of only a methylene group, CH_2:

methyltrimethoxysilane	$CH_3Si(OCH_3)_3$
ethyltrimethoxysilane	$CH_3-CH_2-Si(OCH_3)_3$
n-propyltrimethoxysilane	$CH_3-CH_2-CH_2-Si(OCH_3)_3$
i-butyltrimethoxysilane	$CH_3-CH_2-CH-Si(OCH_3)_3$
	$\qquad\qquad\quad CH_3$

In this series the vapor pressures decrease in accordance with molecular weight, although not in equal steps, from 31 mm Hg for MTMOS, to 14,

Figure 2.2

Evaporation of alkyltrimethoxysilanes.
Methyltrimethoxysilane has significant losses
by evaporation and eventually gels after
about 13 days. Ethyltrimethoxysilane has
less evaporation and gels in about 10 days.
Both *n*-propyl- and *i*-butyltrimethoxysilane
remain as liquids (note the nearly straight
slope of mass vs. time) and eventually completely evaporate leaving no gel behind.

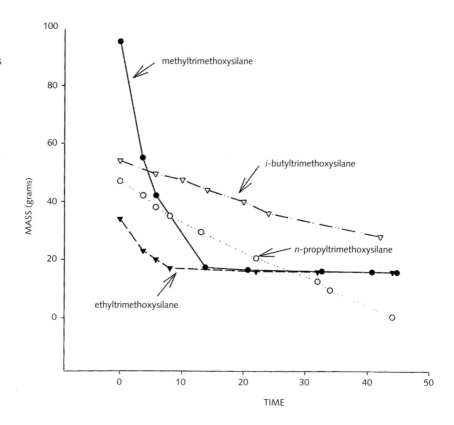

12, and 6 mm Hg for *i*-BTMOS. As the vapor pressure decreases so does
the rate of evaporation. Thus initially the relative rates of evaporation
are MTMOS > ETMOS > *n*-PTMOS > *i*-BTMOS (see fig. 2.2).

Once gelation occurs, the slopes of the curves change dramatically
—they are nearly flat—indicating that evaporation nearly ceases.[12]
These breaks in slope never occur for *n*-PTMOS and *i*-BTMOS because
they never gel: they remain liquids, slowly evaporating, until nothing
remains. This series of alkyltrimethoxysilanes indicates that in the balance between steric hindrance and vapor pressure, steric hindrance takes
on greater importance when the size of the alkyl groups reaches propyl,
that is, *n*-propyltrimethoxysilane. With this compound and the more sterically hindered *i*-butyltrimethoxysilane, the bulk of the alkyl groups prevents the attack on silicon by water that is required to initiate hydrolysis.
The absence of this reaction allows these compounds to evaporate completely. On the other hand, steric hindrance is limited enough for both
MTMOS and ETMOS to allow them to hydrolyze, condense, and gel.
MTMOS reacts more quickly than does ETMOS, but it also evaporates
more quickly and leaves 13 percent solids after 120 days. That ETMOS
deposits 40 percent solids after the same 120 days can be accounted for
by its lower vapor pressure: 14 mm Hg versus 31 mm Hg for MTMOS.[13]
For these two compounds the difference in vapor pressure is much more
important than the difference in steric hindrance between the methyl and
ethyl groups.[14]

As for inductive effects, table 2.4 demonstrates that they also play
a role in the conditions employed here: open containers of neat liquids

reacting with atmospheric moisture at neutral pH. Comparing methyltrimethoxysilane and tetramethoxysilane helps to illustrate the effect. These compounds differ only by the additional oxygen (underlined) on the fourth methoxy group:

$$CH_3Si(OCH_3)_3 \qquad CH_3\underline{O}Si(OCH_3)_3$$

An inductive effect is expected with this *methyl*-for-*methoxy* substitution, increasing the rates of hydrolysis and condensation and, consequently, the amount of gel formed. True to form, MTMOS deposits 13 percent gel and TMOS less than 1 percent, despite its much lower vapor pressure (14 mm Hg vs. 31 mm Hg).[15] The same inductive effect can be invoked when comparing ETMOS with TMOS. These compounds have nearly identical mass and vapor pressure, yet ETMOS leaves 40 percent solids and TMOS almost none. For *n*-propyltrimethoxysilane, only one methylene group larger than ethyltrimethoxysilane, steric effects dominate over inductive effects, and all of the *n*-propyltrimethoxysilane evaporates.

Taken together, the data in table 2.4 help to explain why so few alkoxysilanes qualify as stone consolidants. Alkyl groups larger than *ethyl* create compounds that, without assistance, react too slowly due to steric hindrance. These restrictions leave only methyl- and ethyltrimethoxysilane. Assistance is also required to produce gels from tetramethoxysilane and all of the ethoxysilanes: methyltriethoxysilane, ethyltriethoxysilane, and tetraethoxysilane.

Addition of Water

Assistance to produce gels from tetramethoxysilane and all of the ethoxysilanes may come in the form of liquid water. The data in table 2.4 were generated by exposing alkoxysilanes only to atmospheric moisture, that is, the ambient relative humidity in the laboratory that imparts a relatively low rate of reaction. That rate can be increased by adding liquid water to alkoxysilanes, but because the two are immiscible, a solvent such as ethanol is necessary to form a homogeneous solution.[16] Figure 2.3 is the miscibility diagram for TEOS, water, and ethanol.

The amount of water relative to an alkoxysilane such as TEOS is often expressed as *r*, the molar ratio of water to TEOS. Up to a point, increasing *r* decreases the time it takes for a gel to form. However, as more water is added the solution becomes more and more dilute in TEOS, and the gel time increases. Figure 2.4 illustrates this trend for acid catalyzed solutions of water, TEOS, and ethanol with molar ratios of water:TEOS ranging from 1:1 to 16:1. At 1:1 the gel time was 50 hours; at 2:1, 40 hours; at 4:1, 45 hours; at 8:1, 90 hours; and at 16:1, 105 hours (Klein 1985). The dilution also decreases the amount of gel that forms.

We can infer from the need to add alcohol to provide miscibility with water that alkoxysilanes are hydrophobic. Tetraalkoxysilanes such as TMOS and TEOS in contact with liquid water become less hydrophobic as hydrolysis proceeds due to the formation of water-soluble species

Figure 2.3

Miscibility ternary diagram of tetraethoxysilane, water, and ethanol (Brinker and
Scherer 1990:109). Note that more than
two-thirds of this diagram indicates *immiscibility* among the three components. For
the point shown on the diagram, 58%
ethanol is required to bring 20% water into
miscibility with 22% TEOS.

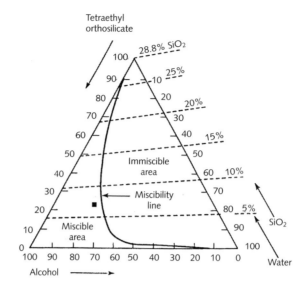

Figure 2.4

Viscosity curves for solutions of
water:TEOS:ethanol, with molar ratios of
water to TEOS ranging from 1:1 to 16:1
(L. Klein 1986). As the molar ratio of
water:TEOS increases from 1:1 to 2:1, the
gel time—indicated by the sharp rise in
viscosity—at first decreases due to the
more rapid reaction resulting from the
additional water. As more water is added
and the solution is more dilute, the gel
time increases.

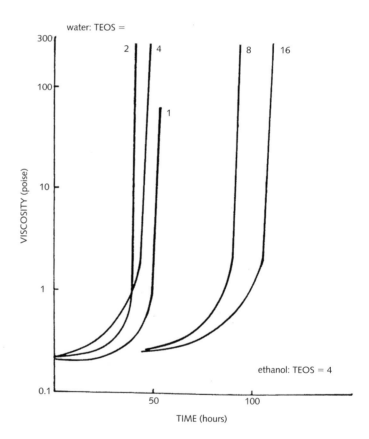

such as $Si(OH)_4$. Miscibility improves for water and unhydrolyzed
TMOS as alcohol is produced during this same hydrolysis:

Hydrolysis $Si(OCH_3)_4 + 4H_2O => Si(OH)_4 + 4CH_3OH$
   hydrophobic   *water soluble mutual solvent*

If sufficient alcohol is generated, the mixture of water and alkoxysilane
will eventually homogenize.

Unlike tetraalkoxysilanes, alkyltrialkoxysilanes such as MTMOS
and MTEOS remain hydrophobic because the methyl groups directly

attached to silicon are unaffected by hydrolysis and condensation. Consequently, no water-soluble silicon compounds are generated by these reactions. However, even with these hydrophobic reaction products, if enough alcohol is generated by hydrolysis, the solution will homogenize in a manner similar to tetraalkoxysilanes:

$$CH_3Si(OCH_3)_3 + 3H_2O => CH_3Si(OH)_3 + 3CH_3OH$$

hydrophobic   *hydrophobic*   *mutual solvent*

As the size of that alkyl group increases, so does its power of water repellency. Larger quantities of alcohol are required to create homogeneous solutions with water, and the resulting solutions are dilute in alkoxysilane. The steric hindrance of larger alkyl groups also limits hydrolysis and condensation such that many of these solutions will not gel. For example, a 10:1:7 mixture of water: *i*-BTMOS:ethanol leaves only 5.4 percent of a thick oil that never gels.[17] These data have the following relevance for stone consolidants. We may wish to take advantage of the slower evaporation of *i*-BTMOS derived from its lower vapor pressure to deposit more gel. Realizing that it will evaporate entirely unless its hydrolysis is assisted, water and alcohol are added. However, this strategy is defeated by the water repellency of the alkyl group that needs more alcohol to make homogeneous solutions with the water that will drive the reaction. In addition, the steric hindrance of the alkyl group does not allow enough hydrolysis and condensation to produce a gel, so that only a thick oil results.

Catalysis

The choice of alkoxysilane and the addition of water affect the rate (and extent) of hydrolysis, but catalysts can exert an even greater influence. Acids and bases promote more rapid and complete hydrolysis than do neutral conditions. More important, the nature of the gel that ultimately forms is determined by the type of catalyst.

Acid catalysis

Hydrolysis occurs by nucleophilic attack on the silicon atom by oxygen contained in a water molecule (Khaskin 1952). The first step in acid catalyzed hydrolysis is the rapid protonation of an alkoxide group; the protonated oxygen leaves silicon even more susceptible to attack by water (Brinker and Scherer 1990):

acid catalyzed hydrolysis

After protonation, hydrolysis proceeds rapidly: for TEOS and $r > 4$ the solution contains mostly $Si(OH)_4$ after twenty minutes (Pouxviel et al. 1989). In these conditions hydrolysis is nearly complete at the commencement of condensation. Because the solution is acidic, and, after hydrolysis, consists mostly of $Si(OH)_4$, some of this newly formed $Si(OH)_4$ is also protonated by the acid catalyst:

first step in acid catalyzed condensation

$$Si(OH)_4 \ + \ H^+ \longrightarrow \ (HO)_3 \underset{H^+}{SiOH}$$

The silicon atom of this now protonated (acidic) species is more prone to (basic) nucleophilic attack, as the most acidic reacts with the most basic. The only nucleophile and virtually the only other species present in this early stage of the reaction is an unprotonated $Si(OH)_4$ molecule that reacts with its protonated sibling to create a siloxane bond. The reaction regenerates the catalyst and water, hence the term *condensation reaction*.

second step in acid catalyzed condensation

$$(HO)_3SiOH \ + \ (HO)_3 \underset{H^+}{SiOH} \longrightarrow \ (HO)_3 Si\text{-}O\text{-}Si(OH)_3 + H_2O + H^+$$

silanol   *protonated silanol*   *siloxane bond*

The solution now contains both monomeric and condensed silanols, $Si(OH)_4$ and $(HO)_3Si\text{-}O\text{-}Si(OH)_3$, each of which is a candidate for protonation by the acid catalyst. Which will be chosen? Brinker (1988) indicates that in this next step, the more basic silanol is protonated according to the model presented in figure 2.5.

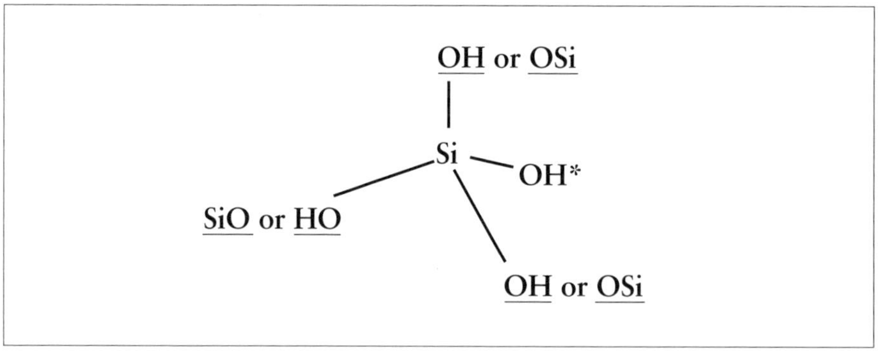

Figure 2.5

Selection for protonation in acid catalyzed condensation. Acid catalyzed solutions may contain both monomeric and oligomeric silanols, $Si(OH)_4$ and $(HO)_3Si\text{-}(O\text{-}Si)_x$ where $x = 1$, 2, or 3, depending on the degree of condensation. Any of these species can be protonated for further condensation. If the OH* is the reaction site, it will more likely be protonated if the other three sites (underlined) are also OH groups, because these groups are more basic (i.e., are more likely to provide electrons) than are SiO groups. For this reason, monomers are more likely to condense with monomers in acid catalysis.

In this model *monomeric* silanols are more basic than *condensed* silanols, so these monomeric silanols are the preferred sites of protonation. This protonated monomeric silanol again seeks the strongest nucleophile in solution—another unprotonated, monomeric silanol, $Si(OH)_4$—in a repeat of the condensation reaction shown above. The importance of this scheme is that in acid catalyzed condensation, monomers prefer to react with other monomers rather than with condensed silanols (i.e., dimers, trimers, and larger oligomers).

Eventually, of course, all monomers are consumed, and condensation must involve oligomers. The most basic silicon atom in this collection of oligomers will be protonated. Examining a fully hydrolyzed trimer illustrates how the selection for protonation is made. The trimer contains two types of silicon atoms: "end" silicons attached to only one other silicon atom through oxygen and a "middle" silicon (italics) attached to two other silicons; that is, it is more condensed.

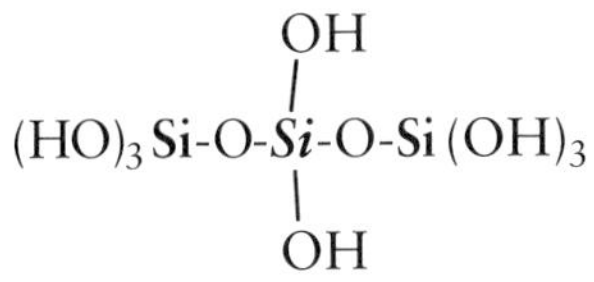

$$\text{(HO)}_3\text{Si-O-}\textit{Si}\text{-O-Si(OH)}_3 \quad \textit{middle silicon}$$

with OH groups above and below the middle silicon.

By the criteria shown above, the less condensed end silicons are more basic than the middle silicon and would more likely be protonated by the acid catalyst. This newly protonated and now acidic silanol condenses with another, more basic silanol, that is, an unprotonated end silanol, to form a new siloxane bond. The central silicon atom is shunned in the condensation reaction:

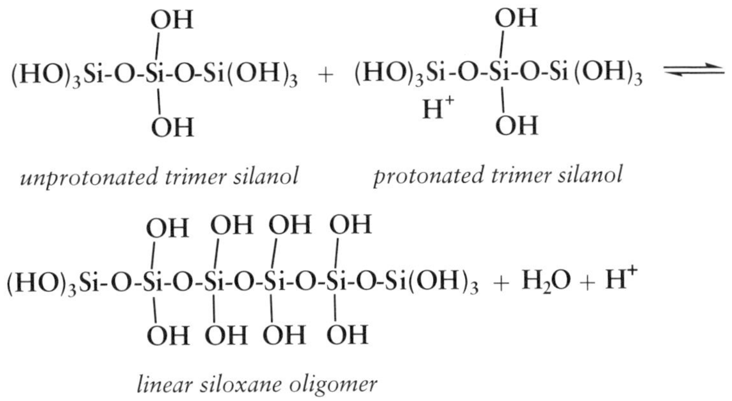

$$\text{(HO)}_3\text{Si-O-Si-O-Si(OH)}_3 \; + \; \text{(HO)}_3\text{Si-O-Si-O-Si(OH)}_3 \; \rightleftharpoons$$

unprotonated trimer silanol   *protonated trimer silanol*

(with OH groups on the silicons, and H$^+$ on the protonated trimer)

$$\text{(HO)}_3\text{Si-O-Si-O-Si-O-Si-O-Si-O-Si(OH)}_3 \; + \; \text{H}_2\text{O} + \text{H}^+$$

(with OH groups above and below the chain silicons)

linear siloxane oligomer

Thus acid catalysis promotes the formation of siloxane bonds by monomer~monomer, monomer~end-group-oligomer, and end-group-oligomer~end-group-oligomer interactions. These conditions produce linear or weakly branched structures.

Base catalysis

Hydrolysis under basic conditions involves the nucleophilic attack of the silicon atom by hydroxyl anion (OH$^-$), followed by displacement of an alkoxide anion (RO$^-$) (Iler 1979; Keefer 1984):

base catalyzed hydrolysis

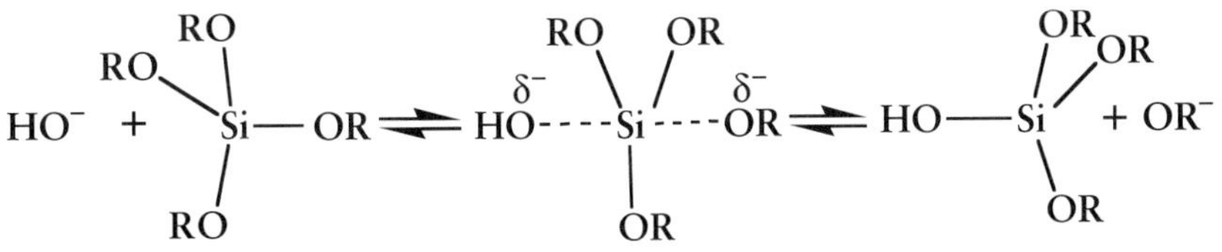

$$\text{HO}^- \; + \; \begin{array}{c}\text{RO}\\\text{RO}-\text{Si}-\text{OR}\\\text{RO}\end{array} \rightleftharpoons \begin{array}{c}\text{RO}\;\;\text{OR}\\ \text{HO}^{\delta-}\text{----Si----OR}^{\delta-}\\\text{OR}\end{array} \rightleftharpoons \begin{array}{c}\text{OR}\;\text{OR}\\\text{HO}-\text{Si}\\\text{OR}\end{array} + \text{OR}^-$$

As with acid catalysis, hydrolysis is rapid in basic conditions. Condensation begins by deprotonation of a monomer silanol by hydroxyl anion to create the silanolate anion (Brinker and Scherer 1990:147; Iler 1979):

first step in base catalyzed condensation

The silanolate anion is itself a strong base or nucleophile and seeks to react with acidic silanols. In the initial stage of the reaction, the most acidic silanols are $Si(OH)_4$. The reaction between the silanolate ion and $Si(OH)_4$ produces a siloxane bond, and the catalyst, the hydroxyl ion, is regenerated (Brinker and Scherer 1990:145):

second step in base catalyzed condensation

The regenerated hydroxyl ion of the base catalyst now seeks the most acidic silanol to deprotonate. Silanols on condensed species, for example, dimers and trimers, are more acidic than those on monomers, and they become the preferred locus for deprotonation by hydroxyl ions. Being more condensed, the middle silanol on a trimer is more acidic than the terminal, or end, silanol and will be deprotonated more quickly by hydroxyl ions.

The preferential activation of condensed silanols in base catalyzed condensation produces more highly condensed structures in solution as compared to the linear or weakly branched structures seen earlier in acid catalysis. The degree of condensation will have significance for the kind of gels that form and their subsequent behaviors.

Gelation, Syneresis, and Ripening

The solutions continue to evolve along the condensation pathways dictated by their catalysts until they form gels. For solutions with $r \geq 2$, the most obvious change as gelation approaches is a rapid increase in viscosity (Debsikdar 1986). The viscosity changes very little until just near the gel point.[18] For solutions with $r < 2$, gelation occurs with less abrupt changes in viscosity (fig. 2.6).

Figure 2.6

Viscosities of sols as they approach and then reach gelation. As shown here, gelation can occur quickly for sols with $r \geq 2$. With $r < 2$ gelation is much less dramatic and can be approached quite slowly with $r = 1$. Acid catalyzed sols with low r values can make good films (Sakka 1984).

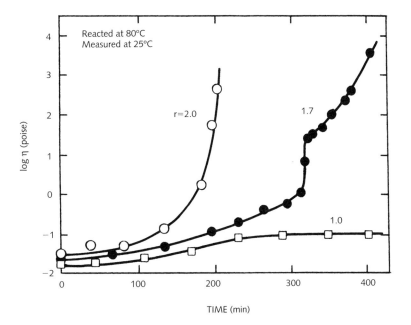

As Brinker and Scherer (1990:358) point out, gelation may be a spectacular event, but it does not represent the end of physicochemical activity for the gel, that is, the solid network and the liquid that permeates it. Condensation (and hydrolysis) continues, and condensation causes the gel to shrink. The shrinkage pushes liquid from the pores in the gel in a process called syneresis.

Other changes occur to the gel without shrinkage. Coarsening or ripening involves dissolution and reprecipitation of parts of the gel. Particles in a gel are interconnected by necks. The convex particles are more soluble than the concave connections, and smaller convex particles are more soluble than larger ones. The net result of these conditions is that smaller particles dissolve and reprecipitate on concave regions such as necks or on larger particles. This process thickens the concave connections, giving the gel more strength and stiffness.

The propensity for ripening depends on pH and is therefore influenced by the original conditions of catalysis. As silicates or siloxane bonds are more soluble at higher pH, ripening is more active with base catalysis that further promotes the creation of more highly condensed species.

Figure 2.7

Tensile stresses created by the stretching pore liquid during drying (after Brinker and Scherer 1990). As the gel begins to dry, pore liquid stretches out to cover the exposed solid gel network. The tension in the liquid can be quite high—up to 100 MPa—and is inversely proportional to the pore sizes.

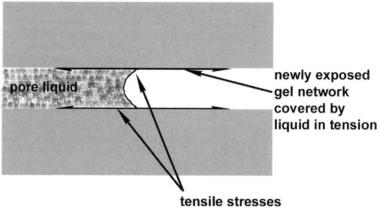

Drying

For gels open to the air, liquid in the pores eventually evaporates and exposes the solid network. The exposure of this network at the surface causes liquid deeper inside the gel to cover this newly exposed solid. In attempting to cover this solid, the liquid stretches, or goes into tension. The tension in the liquid pulls the gel network inward, causing it to shrink further (fig. 2.7). If the liquid cannot flow easily through the solid network from interior to exterior, the liquid in the extremities is more in tension than the liquid in the interior. Under this stronger "pull," the gel at the exterior tries to collapse inward but is restricted by the slower-shrinking gel on the interior. Instead, the gel at the exterior shrinks along its surface and eventually tears or fractures (Brinker and Scherer 1990:493–94).

The likelihood that fracture will occur depends on several factors. First, the question must be asked, how large are the tensile stresses in the liquid? These stresses are inversely proportional to the pore size—that is, smaller pores create larger stresses—and can be as large as 100 MPa, which is sufficient to fracture almost any gel.[19] The second is, how easily can liquid flow through the gel? This is governed by the gel's permeability, which is proportional to the square of the pore size.[20] Brinker and Scherer (1990) report an average pore diameter of 18 Å for an acid catalyzed gel and 125 Å for a base catalyzed gel. Therefore, in acid catalysis, tensile stresses are greater in magnitude because pores are smaller. These same small pores restrict the flow of the liquid and set up a differential tensile stress from the interior to the exterior of the gel.

All other things being equal, it would appear that acid catalyzed gels fracture more readily. However, other factors influence the potential for fracture that depend on catalyst conditions. As described earlier, gels created in basic environments are more condensed and can be both stronger and stiffer than "acid" gels. The strength and stiffness created by the higher degree of condensation are a mixed blessing when it comes to avoiding fracture. Stiffness will resist the shrinkage driven by tension in the liquid, but the greater strength in "basic" gels may not be great enough to resist the tearing forces created by that very same tension. Conversely, acid catalysis creates less condensed, linear or weakly branched gels that are both weaker and more compliant. Flexible gels do not resist shrinkage, and the compliance or flexibility can allow the gel to collapse inward under the tension of the liquid without fracture.

In summarizing this chapter we can now say that the basic reasons for the popularity of alkoxysilanes as stone consolidants are clear, and they concern fundamental properties of alkoxysilanes and their gels: low viscosity for the starting materials and thermal, oxidative, and light stability of the end products thanks to their siloxane backbones. Many compounds form siloxane bonds, but only tri- and tetrafunctional silicon compounds are gel forming, and of these, few have the correct balance of volatility and reactivity and are harmful neither to people nor to stone. Given these limitations, the palette of suitable compounds for stone consolidants is nearly monochromatic, consisting of only a few methoxy- and ethoxysilanes.

The technical nature of the discussion of catalysis, gelation, ripening, and drying may make the connection to stone consolidation seem tenuous at times. However, the gel *is* the consolidating material, and its properties—strength, flexibility, degree of fracture—determine its performance. Also, the *sol* lies within the matrix of the stone as it hydrolyzes, condenses, and gels, and the *gel* lies within this same matrix as it ripens and dries. The composition and structure of the stone may influence each of the important processes—catalysis, gelation, ripening, drying—in the transformation from solution to consolidant, from sol to gel. It is to the interaction between stone substrates and alkoxysilane consolidants that I now turn.

Notes

1. Silicon and oxygen together constitute 74 percent of the earth's crust, largely in the form of siloxane bonds (Press and Siever 1978:14).

2. For an excellent review of terminology, see Grissom and Weiss 1981.

3. More specifically, hydrolysis means the *splitting* of a molecule of water.

4. Other mechanisms of condensation can occur. For example, alcoholic condensation involves the interaction of an alkoxide group and a silanol to form a siloxane bond and alcohol:

   $Si\text{-}OH + CH_3OSi => Si\text{-}O\text{-}Si + CH_3OH$

5. For example, the 1998 *Gelest Catalog*, edited by Arkles, contains more than eight hundred reactive silicon compounds: Gelest, Inc., 11 East Steel Rd., Morrisville, PA 19067.

6. Sramek and Eckert (1986) reported using dibutyldiethoxysilanes in consolidant formulations, and Rhone-Poulenc manufactures products used for stone consolidation that contain phenyltriethoxysilane.

7. A certain amount of cross-linking of linear polymers will create a solid network. At least some tri- or tetralkoxysilanes must be present for this to occur. A good model for this phenomenon is RTV (room temperature vulcanizing) silicone resins. These resins are commonly used as mold-making materials and consist of linear, hydroxy-terminated, dimethylsilicone polymers; a cross-linking agent, tetraethoxysilane; and a catalyst, historically, chemicals such as

dibutyltindilaurate. The action of the cross-linking agent (TEOS) causes the mold material to harden. Two of these components, TEOS and dibutyltindilaurate, are identical to those found in the stone consolidant called Wacker or *Conservare* OH.

8. Arkles (1998) indicates a vapor pressure of 530 mm Hg for SiH_4 at -118°C.

9. Arkles (1998) indicates a vapor pressure of 194 mm Hg for $SiCl_4$ at 20°C and 515 mm Hg for SiF_4 at -100°C.

10. Ethyltrimethoxysilane and ethyltriethoxysilane might also be employed as consolidants but have not been used to date.

11. Toluene, when used as a solvent for resins such as ACRYLOID B72, may take a few hours, days, or even months to depart. In films of ACRYLOID B72 toluene is reported to be retained for more than two months (Horie 1987:104). At present toluene is used infrequently in conservation due to health concerns. For use with ACRYLOID B72 it has largely been replaced with mixtures of acetone and ethanol. ACRYLOID B72 is the trade name of an acrylic resin manufactured by Rohm & Haas. It is a copolymer of ethylmethacrylate and methylacrylate.

12. In this graph ethyltrimethoxysilane gels before methyltrimethoxysilane because the initial masses of the two compounds differ by a factor of nearly three.

13. Molecule for molecule, ETMOS will yield a higher mass return than MTMOS, independent of their vapor pressures. For complete hydrolysis and condensation and without evaporation of the alkoxysilanes, the mass return for ETMOS is 54% and for MTMOS 49%. Thus the larger mass of the unreactive alkyl group accounts for this 5% difference. The 27% (40% vs. 13%) difference demonstrated in table 2.4 is far greater than 5% and must be explained by the lower vapor pressure of ETMOS.

14. The rubbery solid left in the case of ethyltrimethoxysilane versus methyltrimethoxysilane results from the slightly larger ethyl group that prevents a closer packing of the polymers making up the gel. The polymers are probably smaller for ETMOS, and that also imparts more flexibility to the solid.

15. If this were simply a steric effect from the larger methoxy group we would expect a similar rate of reaction and percent mass return as for ethyltrimethoxysilane, which has nearly the same size and vapor pressure as TMOS. But ETMOS deposits even more gel than MTMOS; the difference in the amount of gel deposited for these compounds versus TMOS is explained by the inductive effect of the alkyl groups.

16. When water is added, the rate of reaction increases, not because steric hindrance for any given alkoxysilane is reduced—this hindrance remains the same with or without ethanol—but because more collisions between water and alkoxysilane molecules take place than when atmospheric water is employed.

17. These data were generated in contract research for the Getty Conservation Institute.

18. Arnold (1978) considered this property useful in stone consolidants because the low viscosity allowed penetration into a stone's porous network right up to the gel point. Gelation would then prevent both further penetration and reverse migration back to the surface.

19. The pressure, P, is defined by the following equation: $P = -2\gamma_{LV}\cos\theta/r$ where γ_{LV} is the liquid-vapor interfacial energy, θ is the contact angle between the liquid and the solid, and r is the radius of the pore in the gel (Brinker and Scherer 1990:414).

20. Permeability is defined as $D = (1-\rho)a^2/4f_sf_t$ where $1-\rho$ is the porosity, f_s and f_t are constants related to the character of the pores, and a is the radius (assuming the pore is circular) (Brinker and Scherer 1990:421).

Chapter 3
Influence of Stone Type

In his concluding discussion of the treatment of adobe with ethyl silicate, Chiari (1987:25) states that "the medium to be consolidated plays an active role" in the treatment process and its results. Chiari's observation applies equally to stone. In this chapter I examine in what ways, to what extent, and under what conditions stone type influences the process and results of consolidation with alkoxysilanes.

"Stone type" is employed here in the geological sense, for example, granite, sandstone, limestone, and marble. These names carry information about a rock's composition and structure. It will be shown that mineral composition, porosity (i.e., total accessible porosity and size distribution), and grain size and shape affect the outcome of consolidation.

Mineral Composition

The initial examination of the influence of mineral composition divides the rock kingdom into carbonate rocks such as limestone and marble, consisting primarily of the mineral calcite; and silicate rocks such as granite, sandstone, and gneiss, consisting essentially of the minerals quartz and feldspar. Phyllosilicate minerals such as clays and micas weave their way through all of these rocks and are examined separately below. Charola, Wheeler, and Freund (1984) captured the essence of the mineralogy "problem" in their photograph of calcite and quartz powders consolidated with neat methyltrimethoxysilane reproduced here in figure 3.1. The

Figure 3.1
Fine powders of calcite and quartz treated with neat methyltrimethoxysilane. The quartz powder on the right is fully consolidated into a monolith by neat MTMOS.

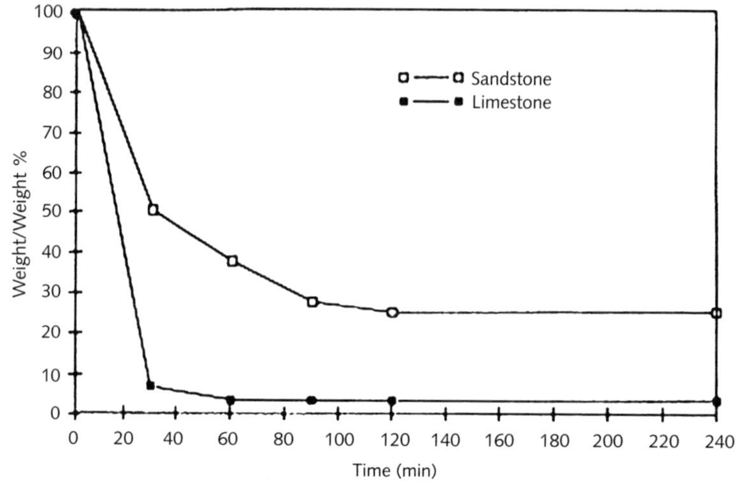

Figure 3.2

Evaporation of neat MTMOS on sandstone
and limestone. MTMOS evaporates almost
completely on Salem limestone (Indiana,
USA) within one hour. Wallace sandstone
(Nova Scotia, Canada) retains 24% MTMOS
and derived products after four hours and
retains that amount after gelation.

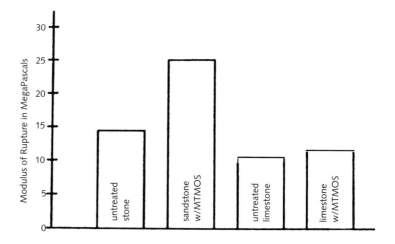

Figure 3.3

Modulus of rupture (MOR) of sandstone
and limestone treated with neat
methyltrimethoxysilane. Wallace sandstone
experiences an 85% increase in MOR;
Salem limestone, only an 11% increase.

fine quartz powder[1] becomes a fully consolidated and rather tough monolith, whereas the calcite powder remains exactly that—a powder.

Moving from powders to rocks, Wheeler et al. (1991) applied
neat MTMOS to cylindrical samples of a purely calcitic limestone
(Salem, Indiana) and a silica-cemented, quartz-rich sandstone (Wallace,
Nova Scotia, Canada). In the first hour after application most of the
MTMOS in the limestone evaporates (fig. 3.2). The sandstone, in contrast, retains 24 percent of the alkoxysilane from two hours and on to
gelation. The near absence of the consolidant in limestone appears to be
confirmed by mechanical testing performed eight weeks later: the sandstone exhibits an 85 percent increase in modulus of rupture (MOR) and
the limestone only 11 percent (fig. 3.3).

Why does so much of the MTMOS evaporate from limestone?
Danehey, Wheeler, and Su (1992) help to provide an answer to this question.
As we see in figures 3.4a, b, and c, their experiments demonstrate that the
polymerization of MTMOS (in solution with water and methanol) is
significantly reduced in the presence of calcite. They tracked this polymerization with silicon-29 nuclear magnetic resonance (^{29}Si NMR). This technique
differentiates "environments" around silicon atoms. The number and type

Figure 3.4a

^{29}Si NMR spectra of a 2:1:2 molar mixture of water (D$_2$O), MTMOS, and methanol. Few monomers (T^0 region) remain after 0.5 hour. After 24 hours, mainly linear (T^2 region) and tri-condensed (T^3 region) silicon atoms are detected.

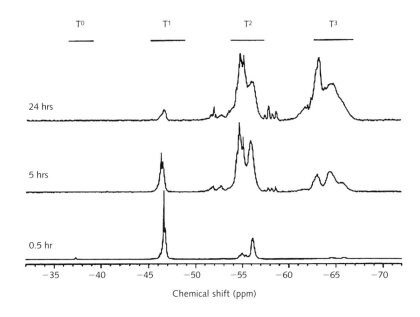

Figure 3.4b

^{29}Si NMR spectra of a 2:1:2 molar mixture of water (D$_2$O), MTMOS, and methanol mixed with powdered quartz. With powdered quartz present, the reaction is similar to the solution with no minerals present.

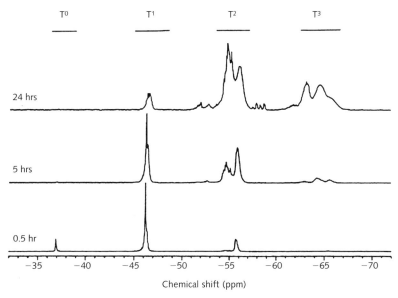

Figure 3.4c

^{29}Si NMR spectra of a 2:1:2 molar mixture of water (D$_2$O), MTMOS and methanol with powdered calcite. Polymerization slows dramatically in the presence of calcite. Even after 24 hours, significant amounts of monomers (T^0 region) remain in solution.

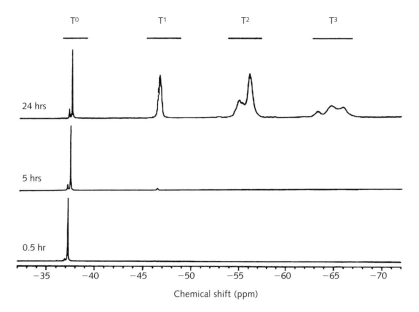

of atoms or functional groups attached to silicon determine these environments. For MTMOS there are four main groups or environments, each designated by a "T" with superscript. T^0, also called the monomer group, represents silicon atoms attached to no other silicon atom through oxygen, that is, no Si-O-Si bonds; T^1, the dimer or end group, indicates silicon atoms attached to one other silicon atom through oxygen; T^2, or the linear group, has two such bonds; and T^3, or the cross-linked group, has three siloxane bonds.[2]

Armed with this information, we can now attack the NMR spectra in figures 3.4a, b, and c. Figure 3.4a shows the spectra at 0.5, 5, and 24 hours for a solution containing 2 moles of water, 1 mole of MTMOS, and 2 moles of methanol. After 0.5 hours the solution contains a range of silicon types: a very small amount of T^0s, or monomers, are present, abundant T^1s and T^2s, and some T^3s. After 5 hours, the monomers have been consumed, and significant increases in T^2s and T^3s are seen, a trend that continues with the 24-hour spectrum. Figure 3.4b shows the NMR spectra for the same solution in the presence of powdered quartz. These spectra are similar to the previous group in Figure 3.4a but with a slight slowing of the reaction represented particularly by the lesser amount of T^3s. With powdered *calcite* a different picture emerges (fig. 3.4c): monomers, or T^0s, are abundant throughout the 24-hour period of analysis. In fact, at 0.5 and 5 hours they are virtually the only species present. Clearly, the overall polymerization of MTMOS dramatically slows down in the presence of calcite.

The x-axes for figures 3.4a, b, and c are marked in parts per million (ppm) and refer to the chemical shift position, or chemical shift, of a given compound relative to a reference compound whose chemical shift is 0 ppm. This shift indicates that the environment around silicon is different from that of the reference compound and resonates with the NMR signal at a new position on the x-axis. In the calcite mixture the chemical shifts for the monomers fall approximately at −37 ppm. The *unhydrolyzed* monomer −resonates at −41 ppm (Danehey, Wheeler, and Su 1992) so that the monomers in the calcite mixture are partially if not completely hydrolyzed:

hydrolysis products of MTMOS monomer

$$
\begin{array}{ccc}
OCH_3 & OCH_3 & \\
CH_3\underline{Si}\text{-}OH & CH_3\underline{Si}\text{-}(OH)_2 & CH_3\underline{Si}\text{-}(OH)_3 \\
OCH_3 & &
\end{array}
$$

As there are abundant hydrolyzed monomers, we can conclude that calcite slows the *condensation* or silicate-bond forming reactions more than the hydrolysis or silanol-forming reactions. This attenuation of condensation promotes evaporation of the more volatile monomers.[3] Therefore, the weight loss and MOR results for limestone presented in figures 3.2 and 3.3 can be explained by the retardation of the condensation reaction brought about by the calcite making up the limestone.

In work following the model of Charola, Wheeler, and Freund (1984),[4] neat MTMOS was again applied dropwise to powdered quartz and calcite in small Nalgene containers until the powders were saturated. The containers were weighed periodically over the next few weeks. For the quartz powder the mass of MTMOS-derived gel stabilized at 28% w/w, and the powder was fully consolidated. The calcite powder remained unconsolidated, with little gel deposited from the MTMOS. Over the next two months, the calcite powder was treated repeatedly with MTMOS until the mass return was the same as for quartz, but the calcite powder remained unconsolidated, confirming that consolidation is not simply a matter of depositing enough treatment material.

Overcoming evaporative loss can be addressed by other means than the repeated applications described above. First, lower vapor pressure monomers such as tetraethoxysilane can be used (see table 2.3). Second, monomers can be partially hydrolyzed and condensed to create a range of low-molecular-weight oligomers.[5] Third, catalysts can be incorporated that increase the rate of reaction and thereby reduce the time it takes for the solid gel to form. Formulations based on these strategies— separately and in combination—have been used on both sandstone and limestone, resulting in greater increases in MOR for both stone types than with neat MTMOS. Goins, Wheeler, and Fleming (1995) applied several formulations of partially polymerized TEOS to limestone and sandstone samples, including one catalyzed version, *Conservare* OH. The gravimetric profiles for the samples treated with *Conservare* OH are shown in figure 3.5; here (unlike fig. 3.2 for neat MTMOS) the percentage of gel retained on sandstone and limestone is at least similar (36% vs. 30%).[6] However, in figure 3.6 it can be seen that despite similar deposits of gel, the increases in modulus of rupture are much lower for limestone— approximately 60 percent—than for sandstone—an average of 220 percent.

So far I have shown that the condensation reactions for uncatalyzed MTMOS solutions are slower in the presence of calcite, a process that permits the evaporation of active ingredients. When strategies are employed that reduce or eliminate evaporation—lower-vapor-pressure monomers, partial polymerization, and catalysis—increases in MOR can be improved for some limestones: from approximately 10 percent for neat MTMOS (i.e., monomeric and uncatalyzed) up to 90 percent for

Figure 3.5

Gravimetric profile of *Conservare* OH on sandstone and limestone. With a TEOS-based, catalyzed consolidant such as *Conservare* OH, the mass return is similar for sandstone and limestone.

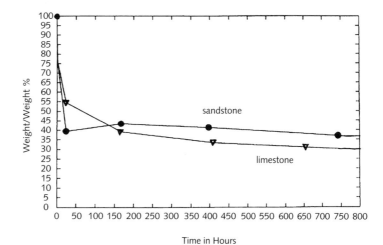

Figure 3.6

Percent increase in MOR of sandstone and limestone treated with TEOS-based consolidants, including *Conservare* OH (Goins, Wheeler, and Fleming 1995). Samples of Ohio Massillon sandstone and Monks Park limestone were treated with seven consolidants based on TEOS. In each case the sandstone experiences a much higher increase in MOR. (TEOS-derived sols were evacuated and then redissolved in the solvents shown along the horizontal axis of this bar graph before application to the stone samples.)

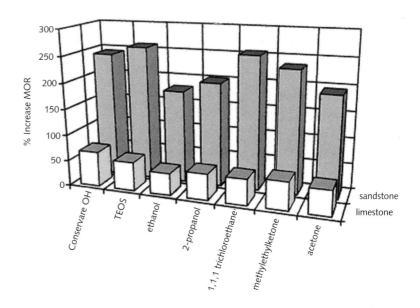

Conservare OH (partially polymerized and catalyzed TEOS). The two limestones tested (Monks Park, Bath, England; and Salem, Indiana, USA) experience much lower increases in MOR than the two quartz-rich sandstones used in the testing (Ohio Massillon, USA; and Wallace, Nova Scotia, Canada).[7] For the catalyzed consolidant the amount of gel deposited is similar for sandstone and limestone, so differences in the amount of gel deposited cannot explain differences in performance. A question still to be answered is whether the gels, that form in contact with different minerals are in all cases the same.

As discussed in chapter 2, acid catalyzed alkoxysilane sols yield different structures from those produced by base catalysis: acid catalysis leads to less condensed "polymeric" gels, and base catalysis to more condensed "particulate" gels. Quartz and calcite are generally considered to have "acidic" and "basic" surfaces, respectively (Fowkes 1987; Goins, Wheeler, and Fleming 1995), and, in contact with reacting alkoxysilane solutions, these minerals could produce more or less condensed gels that in turn may have different consolidating abilities.

Laurie provided the first known observations[8] of the influence of mineralogy on the nature of alkoxysilane-derived gels:

> I have now found that if the silicic ester [i.e., ethyl silicate or partially polymerized tetraethoxysilane] is slightly acid before it begins to hydrolyze, the hydrated silica which is deposited forms a hard glassy layer that constitutes an excellent preservative or cement layer within the pores and on the surface of the stone. If the silicic ester is slightly alkaline, however, the hydrated silica is deposited as a soft gelatinous precipitate which is useless as a cement or preservative.

Limestones and calcareous sandstones are generally
sufficiently alkaline to render the silicic ester alkaline and
so to make the precipitate soft and useless. (1926:1)

From intuition and experience Laurie understood that acid and
base catalysis produce gels of different structures. Goins made similar
observations nearly seventy years later for uncatalyzed MTMOS reacting
in the presence of sandstone, marble, and limestone (Goins 1995; Goins
et al. 1996). She qualitatively assessed the gel's mechanical properties
after gelation. They ranged from "very sticky, formed long tendrils"
when touched with a glass rod for both a control solution and on sandstone to "very weak gels—almost no cohesive strength"—on limestone
and marble (1995:188). The gels that form on the silicate mineral substrates appear more suitable as consolidants than those gels formed on
the calcium carbonate substrates. These observations are consistent with
the MOR testing that shows higher increases for sandstone (Wheeler,
Fleming, and Ebersole 1992).

However, the condition of the gel is less influenced by the mineralogy of the surrounding stone for *catalyzed* formulations. Solid state
^{29}Si NMR was performed on gels derived from *Conservare* OH in the
presence of calcite and quartz and with no minerals added. For a tetrafunctional alkoxysilane like the TEOS found in *Conservare* OH, the
letter Q is used in NMR spectra to designate the different number of
siloxane linkages it may form: Q^0 = no siloxane linkages (monomers);
Q^1 = one siloxane linkage (dimers or end groups); Q^2 = two linkages
(linear groups); Q^3 = three linkages; and Q^4 = four linkages or fully
condensed silicon atoms (Brinker and Scherer 1990).

The spectra in figure 3.7 consist mainly of Q^3 and Q^4 groups. What
is especially striking is their similarity: the minerals have little influence
on the bond distribution in the gel. If the gels from catalyzed TEOS formulations in contact with different minerals are nearly identical, how do

Figure 3.7

^{29}Si NMR spectra of *Conservare* OH
reacted in the presence of no minerals,
quartz, and calcite. The bond distributions
for *Conservare* OH gels that form in the
presence of no minerals, quartz, and calcite are almost identical. Spectra courtesy
Alex Vega

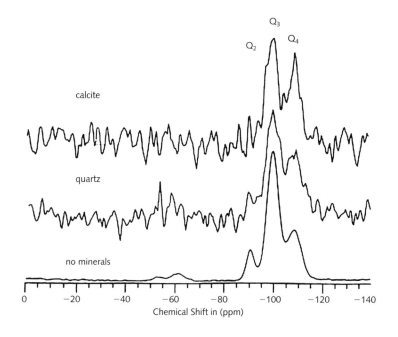

we explain the large differences in performance on the limestones and sandstones shown in figure 3.6?

The stone conservation literature is replete with diagrams similar to figure 3.8 showing linkages between alkoxysilane-derived gels and silicate minerals such as quartz. But do bonds actually exist under the conditions that stone consolidants are normally applied and allowed to react and cure?

Elfving and Jäglid (1992) studied bonding between mineral surfaces and specially chosen alkoxysilanes. They employed minerals commonly found in granite, sandstone, limestone, and marble such as quartz, feldspars (albite and microcline), micas (biotite and muscovite), calcite, and gypsum (the latter is the most common alteration product of calcite in contact with sulfuric acid rain or SO_2 dry deposition). By using *trimethyl*methoxy- and *trimethyl*ethoxysilane, they limited the reaction products to monomers and nonreactive dimers:

(1) $(CH_3)_3SiOCH_3 + H_2O => (CH_3)_3SiOH + CH_3OH$

(2) $(CH_3)_3SiOH + HOSi(CH_3)_3 => (CH_3)_3Si\text{-}O\text{-}Si(CH_3)_3 + H_2O$

(3) $(CH_3)_3SiOCH_3 + HOSi(CH_3)_3 => (CH_3)_3Si\text{-}O\text{-}Si(CH_3)_3 + CH_3OH$

(4) $(CH_3)_3SiOH + HO\text{-}[mineral] => (CH_3)_3Si\text{-}O\text{-}[mineral] + H_2O$

(5) $(CH_3)_3SiOCH_3 + HO\text{-}[mineral] => (CH_3)_3Si\text{-}O\text{-}[mineral] + CH_3OH$

The reaction of the alkoxysilane takes place in the presence of each mineral in the form of a powder. After the reaction, excess liquid is removed by filtration and the powders evacuated. Evacuation further removes excess starting materials—water, monomer, and solvent—and reaction products not attached to mineral surfaces—hydrolyzed monomers, dimer, and alcohol (methanol for the methoxysilane and ethanol for the ethoxysilane). Analysis of the treated mineral powders by infrared spectroscopy determines if the alkoxysilane has attached to the mineral surface by detection of C-H bonds from the methyl groups attached to silicon, as seen above in reactions 4 and 5.

Figure 3.8

Alkoxysilane-derived gel linked to quartz through silicate bonds. The stone conservation literature and product literature for alkoxysilane consolidants often show these silicate linkages between silicate mineral substrates and alkoxysilane-derived gels.

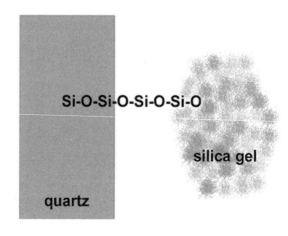

The spectra for the silicate minerals quartz, albite, muscovite, calcite, and gypsum reacted with trimethylmethoxysilane are shown in figure 3.9. C-H bonds (see asterisks in the spectra) are detected for each of the silicate minerals, but, surprisingly, they are in greater abundance for quartz and albite than for muscovite, which contains significantly more terminal OHs to condense with the alkoxysilane.[9] Also, in all of the spectra of silicate minerals, the acid catalyzed solution deposited more alkoxysilane, probably because it is more hydrolyzed under these conditions.

For the nonsilicate minerals, no C-H bonds are detected for gypsum. Calcite, unfortunately, absorbs infrared radiation in the region where C-H bonds are detected. However, if the untreated, base catalyzed, and acid catalyzed spectra are examined closely (see enlargement on the lower right of fig. 3.9), three spectra are identical: no additional absorbances, even small ones, are noted. Therefore, no alkoxysilane has been deposited on the treated calcite.

Elfving and Jäglid (1992) conducted similar experiments with monomeric TEOS and a similar suite of minerals. In this case, the interpretation of the spectra is more difficult because the reaction of TEOS can form products that are not removable by solvent washing and evacuation.

Figure 3.9

These transmission infrared spectra are of minerals alone and reacted with solutions of trimethylmethoxysilane (Elfving and Jäglid 1992). A = untreated minerals, B = minerals and base catalyzed solutions, C = minerals and acid catalyzed solutions. Asterisks indicate the position of the C-H bonds. With each of the silicate minerals (quartz, albite, and muscovite) C-H bonds are noted. With nonsilicate minerals (calcite and gypsum) no C-H bonds are noted. In the lower right is an enlargement of the calcite spectra.

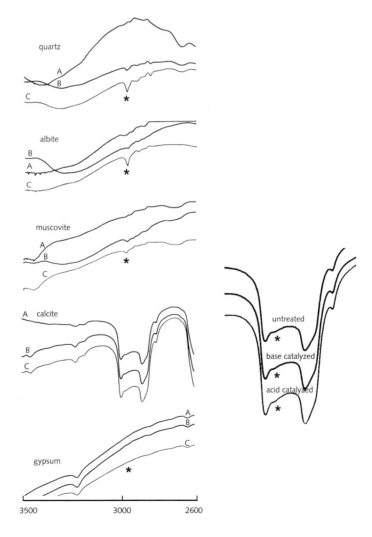

Figure 3.10

These transmission infrared spectra are
of minerals reacted with solutions of
tetraethoxysilane (Elfving and Jäglid
1992). For all silicate minerals (quartz,
albite, microcline, biotite, and muscovite)
C-H bonds are detected (asterisks).
Gypsum shows both C-H bonds and
Si-O-Si linkages (dashed line) that are
probably the result of physisorbed reaction
products of TEOS and not of bonding
between gypsum and TEOS. All such
bonds are absent in calcite.

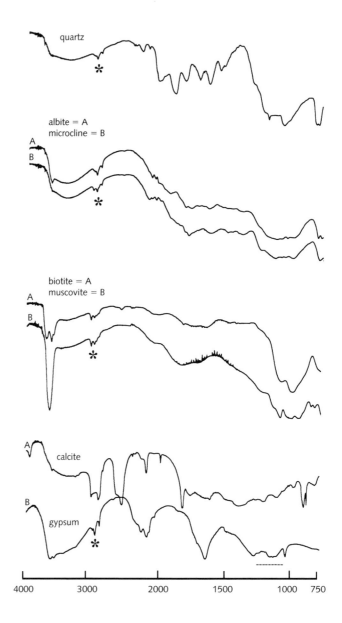

The products can be manifest in the infrared spectra by the detection of
C-H bonds from residual ethoxy groups and by silicate linkages, Si-O-Si
(approx. 1000–1250 cm^{-1}). With quartz, the feldspars albite and microcline, and the micas biotite and muscovite, all treated with TEOS (fig.
3.10), absorbances are seen in the C-H region (asterisks). Unlike the
spectra with trimethylmethoxysilane, TEOS-derived deposits are clearly
present in the gypsum spectrum: C-H (asterisk) and Si-O-Si (dashed
line) absorbances are both present. Therefore, it cannot be said that the
C-H absorbances represent TEOS-derived species *bonded* to the gypsum.
Given that such linkages are absent in the spectra for trimethylmethoxysilane reacted in the presence of gypsum, they are *unlikely* to form with the
bulkier and more sterically hindered TEOS monomer or oligomers. (In
fact, based on this evidence, Elfving and Jäglid concluded that the reaction
products of TEOS are physisorbed on gypsum and not bonded to it.) For
calcite, neither C-H nor Si-O-Si bonds are noted in the infrared spectrum,
indicating that the TEOS-derived products did not bond to or deposit on
the calcite.

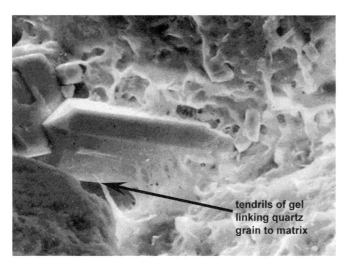

tendrils of gel
linking quartz
grain to matrix

gel does not conform
to the substrate and
is not well linked to it

Figure 3.11a (left)

Scanning electron micrograph of sandstone treated with methyltrimethoxysilane (MTMOS). The MTMOS gel conforms to the surface of the minerals in sandstone and exhibits tendrils of gel linking quartz grains to the matrix. Photo: Elizabeth Goins

Figure 3.11b (right)

Scanning electron micrograph of limestone treated with methyltrimethoxysilane (MTMOS). The MTMOS-derived gel does not conform to the mineral surfaces in limestone and exhibits no linkage to the substrate. Photo: Elizabeth Goins

Another technique frequently used to examine deposits of alkoxysilane in stone is scanning electron microscopy (SEM). This technique can reveal the overall appearance of the gel as well as its relationship to the substrate. Goins (1995) shows differences in MTMOS-derived gels in sandstone and limestone (see figs. 3.11a, 3.11b). Figure 3.11a shows a nearly continuous coating of well-conforming gel with several tendrils attached to a grain of quartz in sandstone. The gel deposit in limestone is, however, isolated and discontinuous: it does not spread over mineral grains and does not bridge adjacent grains or bind grains to the surrounding matrix. (Although these differences are generally noted, Charola et al. [1984] showed tendrils attached to grains in an oolitic limestone.)

The general trend described above has also been observed for TEOS-based consolidants. De Witte, Charola, and Sherryl (1985) provide an excellent view of what they call the "spongy appearance" of the gel on Lavoux limestone (fig. 3.12), and Charola and Koestler (1986) show the classic appearance of TEOS gels in sandstone (see fig. 3.13) that is

Figure 3.12

Scanning electron micrograph of TEOS-derived gel on Lavoux limestone. The gel on Lavoux limestone appears "spongy." Photo: Richard Sherryl

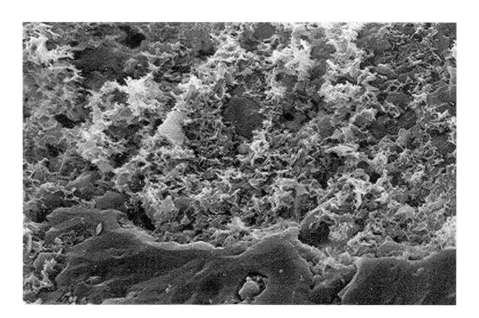

Figure 3.13

Scanning electron micrograph of sandstone treated with TEOS. The gel conforms to the surface and shows a crack pattern typical for OH gels. Photo: Robert Koestler

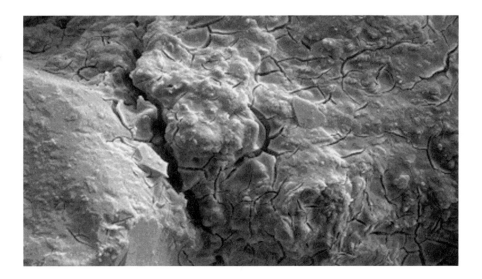

repeated throughout the stone conservation literature. The gels fully conform to their substrates and exhibit shrinkage cracks similar to drying mud. It should also be pointed out that Grissom et al. (1999) demonstrated conformal films with *Conservare* OH on a clay-bearing lime plaster.

The following is a summary of the discussion so far concerning the interactions of alkoxysilanes with some sandstones and limestones or their primary mineral constituents, quartz and calcite:

- The condensation reactions of uncatalyzed solutions of MTMOS, water, and methanol are slower in the presence of calcite. The reduced rate of condensation allows large amounts of MTMOS to evaporate. Only a slight reduction in the rate of condensation is seen with quartz. With neat MTMOS, evaporation is significant in the presence of calcite powder or a purely calcitic limestone, an effect not noted with quartz powder or a quartz-rich sandstone. It is therefore likely that with neat MTMOS in contact with calcite the rate of condensation is reduced in a manner and degree similar to solutions of water, MTMOS, and methanol. Neither effect—reduction in the rate of condensation or significant evaporation—has been noted for catalyzed, partially polymerized TEOS solutions such as Wacker OH.

- Gels formed from solutions of MTMOS, water, and methanol in the presence of a calcitic limestone are initially weaker and less coherent than are gels from the same solutions in contact with quartz powders or quartz-rich sandstones. In contrast, solid state ^{29}Si NMR shows that gels derived from *Conservare* OH have nearly the same bond distribution in contact with quartz and calcite powders, which suggests that the gels are similar with this catalyzed formulation.

- For quartz and other silicate minerals commonly found in sandstones, solutions of MTMOS, water, and methanol or catalyzed, partially polymerized TEOS formulations create conformal and relatively smooth gel coatings. In some instances bridges are formed between mineral grains or between mineral grains and the surrounding matrices commonly found in sandstones. In addition, bonds form between alkoxysilanes such as trimethyl-methoxysilane, or TEOS, and silicate mineral substrates such as quartz, feldspars, and, to a lesser degree, micas. Conformal films, gel bridges, and bonds are usually absent with calcite powders and some purely calcitic limestones.

- Fine quartz powders are strongly consolidated with neat MTMOS. Fine calcite powders remain *un*consolidated.
- Some quartz-rich sandstones exhibit substantially higher *increases* in modulus of rupture than do some purely calcitic limestones.

The Clay Problem

Clay-bearing stones often require consolidation because of their tendency to deteriorate (Rodriguez-Navarro et al. 1997). It might appear that clays (and phyllosilicates generally), with their abundant OH groups, as shown in the infrared spectrum of kaolin in figure 3.14, offer receptive surfaces for alkoxysilanes. Nonetheless, Elfving and Jäglid's (1992) infrared spectra, shown in figure 3.9, demonstrate that less trimethyl-methoxysilane was deposited on muscovite than on quartz and feldspars, which have lower concentrations of Si-OHs. In addition, neat MTMOS and Wacker OH provide little consolidation to fine powders of illite clay or other phyllosilicates such as chlorite, muscovite, or biotite[10], and clay-filled composites using vinyltrimethoxysilane as the coupling agent had much lower flexure strengths than did similar composites filled with alumina, pyrex, and silica (Plueddemann 1991).[11] Finally, Sattler and Snethlage (1988) state that clay-cemented sandstones exhibit only half the strength increases of silica-cemented sandstones when each is treated with ethyl silicate.[12]

Figure 3.14

Fourier transform infrared (FTIR) spectrum of kaolin clay (kaolinite). Si-OH bonds are abundant in kaolin in the region of 3600 cm⁻¹. Despite the abundance of these hydroxyl groups, kaolin is not well consolidated by alkoxysilanes.

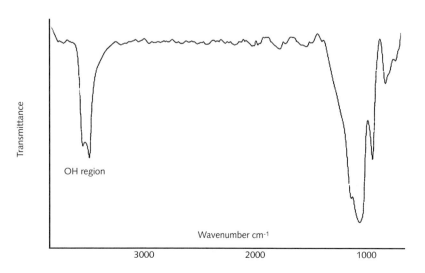

Other studies, however, provide data that apparently contradict these results. Hosek and Sramek (1986) indicated successful treatment of gaize, a cretaceous, marly limestone, with alkylalkoxysilane. Thickett, Lee, and Bradley (2000) obtained 140 percent increases in crushing strength for an 80/20 mixture of sepiolite and palygorskite treated with Wacker OH, and SEM showed conformal films of OH gel similar to those found on sandstone for a sepiolite-containing limestone. Grissom et al. (1999) showed that lime plaster containing from 9 to 15 percent smectite clay (the authors indicate it is probably montmorillonite) had a 355 percent increase in tensile strength when treated with *Conservare* OH, and, again, SEM revealed conformal films of OH-derived gel. Brethane, the catalyzed MTMOS-based consolidant, created a tough monolith from a well-mixed combination (10% w/w) of fine powders of illite clay and calcite but gives little consolidation to calcite by itself. Caselli and Kagi (1995) treated samples of a quartz sandstone with an (unspecified) clay-rich matrix using Wacker OH, Wacker H, and Brethane. In compressive strength tests, treated and dry samples were about 100 percent stronger than those untreated and dry. In wet conditions, treated samples were 150 percent stronger than untreated samples. (For this stone, untreated wet samples are about half the strength of untreated dry samples.)

Other considerations come into play with clay-bearing stones. Felix and Furlan (1994) studied the behavior of several Swiss building stones, including the clay-containing Villarlod molasse.[13] They first looked at the dimensional change of the molasse after treatment with an ethyl silicate consolidant (fig. 3.15). Compared to the clay-free limestone from Neuchatel, the molasse shows considerable shrinkage with ethyl silicate treatment—nearly 1.5 mm/m—that might be accounted for by the preferential deposit of ethyl silicate between the plates of clay minerals noted by Snethlage (1983).

Figure 3.15 illustrates the tendency of the untreated Villarlod molasse to expand with prolonged immersion in water (Felix and Furlan 1994; Wendler et al. 1992). This kind of expansion can lead to deterioration of the stone by only wetting and drying, and Snethlage and Wendler (1991) warned that it may be dangerous to treat clay-bound sandstone with alkoxysilanes because the treatment increases this dilatation.

It is difficult to draw definitive conclusions from the data and information on clay minerals presented above. Tentatively, it can be said that

Figure 3.15

Deformation of Villarlod molasse and Neuchatel limestone with water and ethyl silicate (C. Felix 1995). The top of the graph shows the expansion of Villarlod molasse with the uptake of water; the bottom shows the shrinking of the molasse with ethyl silicate. In the center, the clay-free Neuchatel limestone neither expands with water nor shrinks with ethyl silicate.

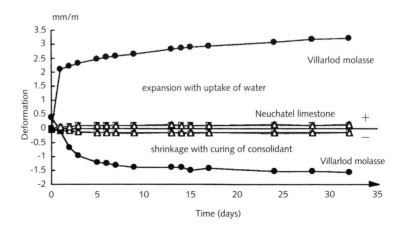

when clays are present in limestones they improve the performance of consolidation with alkoxysilanes when measured by most mechanical tests. Conversely, clays appear to reduce that performance on sandstones. Some clay-bearing limestones can shrink considerably on treatment with ethyl silicates, and some clay-bearing sandstones and limestones are at greater risk to wetting and drying deterioration after treatment. The clays mentioned here are not all of the same structure: sepiolite and palygorskite, illite, montmorillonite, and kaolin are from different clay groups. These structures may have an influence on their ability to be consolidated, but how this influence occurs cannot be determined from the current stone conservation literature.

Other Stone Types

As discussed in chapter 1, alkoxysilanes have been used on other stone types. For volcanic rocks, Tabasso, Mecchi, and Santamaria (1994) showed a 100 percent increase in compressive strength and a 57 percent increase in ultrasonic velocity for a pyroclastic tuff treated with Tegovakon V, an ethyl silicate consolidant. These stones are often characterized by high porosity and low mechanical strength. Useche (1994) studied a volcanic tuff and found that treatments with Wacker H and OH brought about 66 to 296 percent increases in compressive strength, depending on dilution, contact time with the consolidants, and number of applications. For plutonic rocks, Rodrigues and Costa (1996) made an extensive study of the conservation of granites. Treatment of granite with ethyl silicate yielded bending-strength increases similar to sandstone (>170%) and a 74 percent increase in ultrasonic velocity. Treatments have also been carried out with *Conservare* OH on scoria and schist[14], and Jerome et al. (1998) carried out treatments also on a calc-schist.

The most important other stone type that has been consolidated with alkoxysilanes is marble. Marbles are characterized by high carbonate content, most often calcite. They also contain a wide variety of accessory minerals: phyllosilicates such as micas and chlorites, chain silicates such as tremolite, and nonsilicates such as pyrite. Given the dominant carbonate mineralogy, it would be expected that alkoxysilanes perform on marble as they do on purely calcitic limestones, that is, yielding relatively low strength increases, and this is generally confirmed by laboratory testing.[15]

Given the poor or only modest performance of these consolidants on calcite powders and purely calcitic limestone described earlier in the chapter, how do we explain the frequent use of alkoxysilanes on marble? Two factors are relevant in this discussion: the most common form of deterioration of marble is granular disintegration;[16] and, unlike most sedimentary rocks, marble consists of relatively large grains in nearly direct contact; that is, there is no intergranular matrix or cementing material. Consequently, the porosity of marble, even decayed marble, is low, and the shape of the intergranular spaces is sheetlike—large in two dimensions and small in the third dimension (see fig. 3.16).

Figure 3.16

Two views of the structure and inter-granular space in marble. On the left is a graphic representing grains and intergranular space in marble showing its sheetlike structure. The back-scattered scanning electron micrograph on the right also shows the narrow spaces between grains compared to the sizes of the grains themselves and the absence of matrix or cementing material (Lindborg 2000).

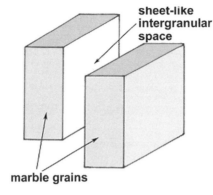
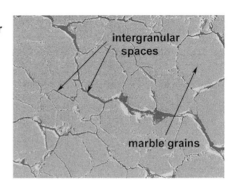

Figure 3.17

Scanning electron micrograph of Lasa marble treated with Wacker OH. This scanning electron micrograph shows the cracked gel from Wacker OH nearly filling the intergranular space of Lasa marble (Ruedrich et al. 2002). Note that the gel conforms well to the relatively flat surface where a large grain was removed.

Alkoxysilane solutions enter these intergranular spaces and later gel and shrink. What the gel looks like in these spaces is clearly demonstrated by Ruedrich, Siegesmund, and Weiss (2002). The SEM shown in figure 3.17 helps to explain why OH-type gels are capable of stabilizing marble against granular disintegration. Although the gel does not *adhere* to this grain, it nearly fills the spaces surrounding the loose grain and prevents it from moving and being dislodged.

Other Issues of Structure

The influence of the structure of marble on its ability to be stabilized against granular disintegration with alkoxysilanes leads naturally to a discussion of issues of structure in other stone types. For example, neat MTMOS creates a monolith with 200 mesh quartz, and the MTMOS gel bonds to the surfaces of the quartz grains. As can be seen in figure 3.18, larger quartz grains, approximately 0.4 mm, remain unconsolidated with

Figure 3.18

On the right are fine grains of quartz that have been consolidated into a monolith with neat MTMOS. The coarser grains on the left remain unconsolidated due to the large and more pocketlike intergranular spaces.

neat MTMOS despite the chemical compatibility between the mineral and the consolidant.

Figure 3.19 helps to illustrate why the coarse grains are not consolidated. For spherical close-packed particles of diameter **d**, the diameter of the smaller sphere lodged in the intergranular space is 0.15 **d**.[17] If **d** is 0.4 mm (i.e., 400 μm), the smaller sphere is 60 μm. For the fine quartz grains (i.e., 70 μm) the smaller sphere is about 10 μm. Therefore, the maximum space that can be bridged by MTMOS gel to create a monolith from quartz grains is between 60 μm and 10 μm. These numbers tend to support Wendler et al.'s (1999) contention that the maximum bridging capability for ethyl silicate gel is 50 μm, which would place the maximum grain size at approximately 325 μm (see also Rolland et al. 2000).

This general theme is also supported by the work of Nishiura (1987a:805), who stated that alkoxysilanes are "suitable for consolidating powdering or chalkingly decayed stone . . . but . . . not for granularly decayed stone." The basis for this statement was the results obtained from evaluating the ability of consolidants to bind granules (1–4 mm in diameter) of the porous tuff Oya stone.[18] SS-101, an MTEOS-TEOS consolidant,[19] failed in a succession of impact and abrasion tests even for the silicate minerals in this volcanic tuff. By Wendler's (1999) yardstick, these particles (1000–4000 μm) are well beyond the ability of alkoxysilanes to bind them together.

Figure 3.19

The intergranular space for spherical particles of diameter d is approximately 0.15 d. For quartz grains of 0.4 mm (400 μm) diameter, which are not consolidated by MTMOS, the intergranular space is 60 μm. For particles of 70 μm, the smaller sphere is approximately 10 μm.

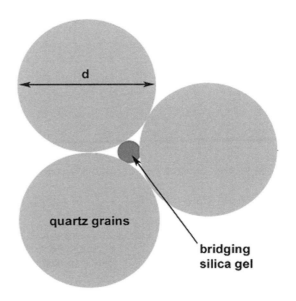

The deposit of alkoxysilane gel in the pore spaces alters the overall porosity, the pore size distribution, and the water transport properties of stone (Fitzner and Snethlage 1982). Hammecker, Alemany, and Jeannette (1992) reported on a dolomitic limestone (Laspra) and a bioclastic limestone (Hontoria), each treated with ethyl silicate (RC70) and a mixture of ethyl silicate and dimethylsiloxane (RC80). After treatment with RC70, the porosity of Laspra was reduced from 30 to 28 percent, and for Hontoria, the bioclastic limestone, virtually no change in porosity was noted. Mercury intrusion porosimetry indicated that for ethyl silicate and Laspra stone, the gel was deposited in the fine pores; and for Hontoria stone, the pore size distribution appeared to be largely unaffected. Perez et al. (1995) also evaluated stone samples from Spanish monuments by mercury intrusion porosimetry. Among the consolidants tested was Wacker OH. For a calcareous sandstone (Puerto de Santa Maria) Wacker OH deposited gel in the pores 20 to 40 µm in diameter and to a lesser degree in pores ranging from 0.2 to 10 µm. A similar trend was noted with a limestone from Granada.

For Lecce stone, a porous quartz-and-glauconite-containing fossiliferous limestone, Tabasso and Santamaria (1985) measured a 38 percent decrease in overall porosity, a 39 percent decrease in water vapor permeability, and a 15 percent increase in the drying index with Wacker OH. On a stone similar to Lecce from a Mayan site in Belize, Ginell, Kumar, and Doehne (1991) noted a 41 percent decrease in porosity with filling of the smallest pores. For decayed Carrara marble treated with Wacker OH and Wacker OH–methylphenylsiloxane, both in sequence or in solution, Verges-Belmin et al. (1991) demonstrated a reduction of pores in the 6 to 16 µm range. Alessandrini et al. (1988) found a general reduction in porosity below 2 µm for both ethyl silicate–dimethylsiloxane (RC80) and ethyl silicate–methylphenylsiloxane (RC90) treatments on decayed Carrara marble. This collection of data, once again, supports the notion that alkoxysilane gels tend to deposit in, and therefore alter the percentage of, the smaller pores (< 50 µm) as measured by mercury porosimety.[20]

Darkening and Color Changes

Another important factor for all consolidants is color change or darkening of treated stone. Darkening and color shifts occur in all porous materials when they imbibe liquids because air/mineral interfaces are replaced by liquid/mineral interfaces. Under these conditions the amount of light that is reflected from the mineral surfaces is reduced because the refractive index of the liquid is higher than that of air. We see this phenomenon when stone becomes wet with rain and appears darker. With consolidants the applied liquid eventually turns to a solid, and if pores are filled or coated the darkening remains.[21] Figure 3.20 shows this effect that has persisted for twenty years in stone treated with an alkoxysilane consolidant.

The degree of color change and darkening depends to some degree on the minerals present in a stone. The effect of two consolidants, RC90 and Wacker OH100, on several minerals is shown in figure 3.21.[22] In all cases, RC90, a catalyzed mixture of ethyl silicate and methylphenyl-

Figure 3.20

Areas of this green sandstone imbibed excess consolidant, which caused darkening in localized areas. After curing, the consolidant has left pores coated or filled, and after twenty years the darkening still remains.

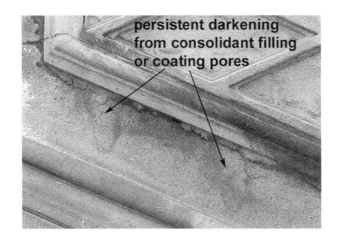

silicone in toluene, resulted in greater yellowing than Wacker OH100, and, in all cases except biotite, RC90 produced more darkening. Light-colored minerals, quartz, albite, and calcite, were much less affected than the dark minerals, red iron oxide, biotite, illite, and clinochlore. Costa and Rodrigues (1996) noted similar trends in their work on granites. Since granites contain relatively large grains of both light and dark minerals, they were able to isolate changes to each after treatment, determining that dark minerals showed a greater darkening effect and light minerals greater yellowing. The effects on several other stones are shown in figure 3.22. Similar to the powdered minerals, the light-colored

Figure 3.21

Mineral powders treated with RC90 and Wacker OH100 consolidants. Darkening is more evident on darker minerals such as biotite, illite, clinochlore, and red iron oxide than on lighter minerals such as calcite, quartz, and albite. Photos: John Campbell. (To view this figure in color, see note on the copyright page.)

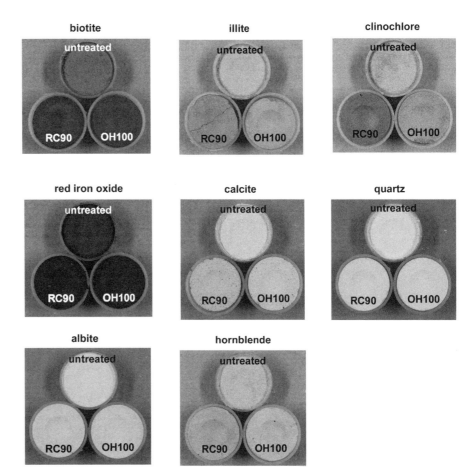

Figure 3.22

Stone samples treated with Wacker OH100 and RC90. The light-colored marble and Abydos limestone show little shift in color with treatment. The tan-yellow stones (Ohio Massillon, Fuentidueña, and Monks Park) become noticeably darker, and the Longmeadow sandstone with its red iron oxides shows an even greater shift. The volcanic rock, the rhyolite from Colorado, also darkens considerably. Photos: John Campbell. (To view this figure in color, see note on the copyright page.)

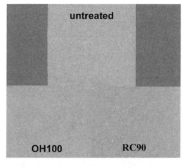

marble (unknown origin)

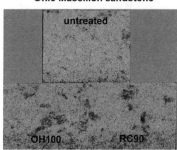

Ohio Massillon sandstone

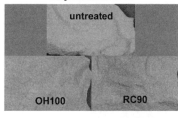

Abydos limestone

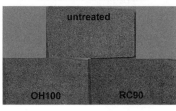

Pietra serena sandstone

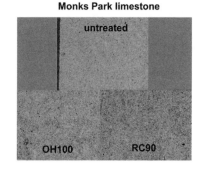

Fuentidueña Apse dolomitic limestone

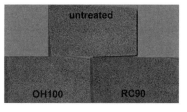

Monks Park limestone

Pietra serena sandstone

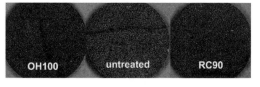

Longmeadow sandstone

Abydos limestone and the nearly white marble are the least affected. With an equally light-colored Lecce stone (not shown in fig. 3.22), Tabasso and Santamaria (1985) reported no color change with Wacker OH, and Duran-Suarez et al. (1995) indicated little or no change for a biocalcaremite. The three tan-to-yellow stones (Ohio Massillon, Fuentidueña, and Monks Park) in figure 3.22 show significant shifts to darkening. One of these is the dolomitic limestone from the Fuentidueña Apse at the Cloisters. The apse has undergone several treatments with *Conservare* OH in the past two decades. Figure 3.23 shows an area of the apse before, one month after, and five years after treatment. The initial change is a dramatic darkening, but after five years the change is almost imperceptible.

The Monks Park limestone and the Ohio Massillon sandstone contain yellow iron oxides that also darken with treatment. The red Longmeadow sandstone exhibits color shifts similar to the red iron oxide powder in figure 3.21, suggesting it is this mineral in the sandstone that is responsible for the darkening. A red scoria object shown in figure 3.24 has an even more pronounced color shift on treatment with *Conservare* OH that required one year to return to the original appearance. Another volcanic rock, the rhyolite from Colorado in figure 3.22, also showed significant darkening with both Wacker OH100 and RC90. With highly porous volcanic tuffs it has been shown that this darkening diminishes with artificial weathering, and with time in an indoor environment (Wheeler and Newman 1994; Nishiura 1986).

Returning to the marble in figure 3.22, there is a slight blue shift for Wacker OH100 and RC90 in these fresh, unsoiled samples.[23] For weathered and soiled Carrara marble, Verges-Belmin et al. (1991) noted a *yellow* shift with RC90 and attributed it to the solvent toluene that has solubilized organic material in black crusts and redeposited them in the pores of the marble.[24]

In summarizing this chapter, we have seen that calcite substrates have a strong influence on neat MTMOS and solutions of water, methanol, and MTMOS. That influence consists in promoting the evaporation of the alkoxysilane by slowing the condensation reaction. Catalyzed formulations overcome the problem of evaporative loss on the consolidant, and this loss is further reduced by using lower-vapor-pressure starting

materials such as TEOS, MTEOS, and their oligomers that typically are present in products such as Wacker OH and Wacker H.

For uncatalyzed formulations, calcite influences the nature of the gel that forms in contact with the mineral. Catalyzed formulations minimize or eliminate the chemical catalytic effect that calcite has on gel formation. Even with catalyzed formulations, however, calcite often alters the physical relationship of the gel to the substrate and the physical nature of the gel itself. In quartz and sandstone substrates, gels are filmlike and, although cracked, conform to the mineral surfaces. In calcite and purely calcitic limestones, gels are either spongy or in isolated patches on the mineral surfaces.

Alkoxysilanes do not bond to calcite, but they do bond to silicate minerals, particularly quartz and feldspars. This bonding (or lack of bonding) is the single most important factor that influences the performance of consolidated stone in mechanical testing: quartz-rich samples experience much higher strength increases than do stone samples that contain only calcite.[25]

Despite its predominantly calcite (or dolomite) mineralogy—often nearly 100 percent—marble can be stabilized against granular disintegration or sugaring with OH-type formulations. The positive effect on marbles results from its structure: large grains, in close proximity, such that pore spaces are sheetlike. The gel deposited in these spaces limits the movement of loose grains and thereby reduces losses by granular disintegration. Pore structure (and the related grain size and shape) can also work against consolidation with alkoxysilanes. Stone with pores larger than 50 μm and consisting of grains that are round and not in close contact will not consolidate well even if the grains are silicate minerals.

Much more work needs to be done on the influence of clay minerals on consolidation. Clays in limestone appear to improve the consolidating ability of alkoxysilanes and even produce conformal films of gel similar to those that appear in sandstone. However, some clay-bearing limestones shrink considerably with treatment. Also, clay-bearing sandstones experience lower strength increases than do similar sandstones without clay.

Finally, there is the issue of darkening of stone treated with alkoxysilanes. Stones most likely to darken are those containing dark minerals, phyllosilicate minerals, and iron oxide minerals, and those with fine pores or fine matrices. Alkoxysilane gels preferentially deposit in fine pores and can fill and coat them and result in darkening. The fine matrices of some sandstones, often made up of clays and iron oxides, will also darken with treatment. The same can be said for the fine ground mass that surrounds larger grains in porphyritic volcanic rocks such as rhyolite.

There is a closing cautionary note: It is risky to make general statements about the performance of alkoxysilane consolidants based on mineralogy or stone type alone. The matter is more complex, as I show in the next chapter.

Notes

1. In this case "fine" means 200 mesh or 0.075 mm.

2. These four groups are represented in more detail below.

Monomer Group, T^0

$$CH_3 \qquad\qquad CH_3 \qquad\qquad CH_3 \qquad\qquad CH_3$$
$$CH_3O\underline{Si}OCH_3 \quad CH_3O\underline{Si}OH \quad CH_3O\underline{Si}OH \quad HO\underline{Si}OH$$
$$OCH_3 \qquad\qquad OCH_3 \qquad\qquad OH \qquad\qquad OH$$

These compounds can also be represented as $(CH_3O)_{(3-x)}\underline{Si}(OH)_x$ $(x = 0,1,2,3)$.

Dimer or End Group, T^1

$$CH_3 \quad CH_3$$
$$(CH_3O)_{(2-y)}\text{-}Si\text{-}O\text{-}\underline{Si}\text{-}(OCH_3)_{(2-x)} \qquad (x \text{ and } y = 0, 1, 2)$$
$$(OH)_y \quad (OH)_x$$

This represents nine different dimers, because x and y vary independently. The hydroxy and methoxy groups in the "y" positions could also be replaced with silicate linkages such that the (underlined) silicon atom detected by NMR would represent an end group.

Linear Group, T^2

$$CH_3 \ CH_3 \ CH_3$$
$$V\text{-}Si\text{-}O\text{-}\underline{Si}\text{-}O\text{-}Si\text{-}X \qquad V, W, X, Z = CH_3O, OH \text{ or } Si\text{-}O$$
$$W \quad Y \quad Z \qquad Y = CH_3O \text{ or } OH$$

This represents many different compounds because V, W, X, Y, and Z vary independently.

Cross-Linked Group, T^3

$$CH_3 \ CH_3 \ CH_3$$
$$W\text{-}Si\text{-}O\text{-}Si\text{-}O\text{-}Si\text{-}Z \qquad W, X, Y, Z = CH_3O, OH \text{ or } Si\text{-}O$$
$$X \quad O \quad Y$$
$$Si$$

Again, this represents many different compounds.

3. The vapor pressures of the monomer silanols are not known because they are difficult to isolate. The reactions here are slightly different from those with neat MTMOS because water has intentionally been added to the system.

4. The experiments described in this paragraph were performed by the author.

5. See table 2.3 for vapor pressures for alkoxysilanes.

6. The theoretical mass return for *Conservare* OH assuming complete reaction and complete evaporation of solvents is approximately 23%. We can therefore assume that some solvent remains, and the reaction is not complete.

7. These stones have similar porosities, 23%, but have somewhat different pore size distributions.

8. These were soon followed by similar observations by King and Threfall (1928), King (1930), Ellis (1935), and Shaw (1945).

9. It is noteworthy that only approximately 6% of the hydroxyl sites on albite, a feldspar, are involved in bonding, even when excess alkoxysilane is present.

10. This confirms the results of Elfving and Jäglid, who reported few bonds for mica.

11. The clay is designated ASP-400 and is probably kaolin.

12. They also caution against making broad generalizations about such treatments. The successful treatment of clay-bearing Egyptian limestones at the British Museum and the Metropolitan Museum of Art has been well documented (Thickett, Lee, and Bradley 2000; Wheeler et al. 1984).

13. Tomkeieff (1983) defines molasse as a feldspathic sandstone with calcareous cement.

14. The scoria was the material for a chair by Scott Burton (see fig. 3.24). Scoria is the crust on lava flows that is heavier, darker, and more crystalline than pumice (Bates and Jackson 1984). The calc-schist was from the wall on Fifth Avenue above Ninetieth Street for Central Park in New York City. The schist was from the John Bartram House in Philadelphia. Jerome et al. (1998) call this a marble in the title, but the description in the text suggests it is a calc-schist.

15. It is surprising to note how little mechanical testing of treated marble can be found in the literature. See Wheeler 1991.

16. This form of deterioration is often referred to as "sugaring."

17. This can be derived by simple trigonometry.

18. For a description of Oya stone, see Wheeler and Newman 1994.

19. SS-101 is an ethyl silicate consolidant produced by COLCOAT, Ltd., Tokyo.

20. This comment should be tempered by the fact that the maximum pore size that can be measured by mercury porosimetry is about 100 µm.

21. As we have seen from scanning electron micrographs and mercury porosimetry, it is more likely that larger pores are coated and smaller pores are filled.

22. This work was performed by John Campbell when he was a graduate student at New York University, Institute of Fine Arts, Conservation Center.

23. Data are from the work performed by John Campbell (see note 22).

24. Color changes of marble have also been attributed to solvents such as toluene that are present in these consolidants. They can solubilize organic material in black crusts and redeposit it in the marble and produce a yellow color.

25. Ruggieri et al. (1991) also stated that ethyl silicate gels break down in travertine.

Chapter 4

Commercial and Noncommercial Formulations

The purpose of this chapter is to discuss the contents of commercial and noncommercial formulations of stone consolidants based on alkoxysilanes. As we saw in chapter 2, those contents produce conditions in the liquid state that help to determine some of the properties of the consolidating gel. They may also determine the conditions under which a given formulation should be applied.

Commercial Formulations

The era of consistent and widespread use of alkoxysilanes for the consolidation of stone began with the patenting of the Wacker products in the early 1970s (Bosch 1973; Bosch, Roth, and Gogolok 1976). Although the patent itself maps out a terrain large enough to encompass the development of future products, the two commercial formulations that most clearly stem from this work are the now-classic OH and H.[1] Versions of these products are still in use today, more than thirty years after the patent, and are by far the most commonly used stone consolidants.

Wacker OH (Ger., Ohne Hydrophobie, i.e., without water repellent) is an ethyl silicate– or tetraethoxysilane-based consolidant. Gas chromatography–mass spectrometry (GC-MS) analysis of the formulation in the early 1980s showed that it contained mostly TEOS monomer and some dimer along with smaller amounts of methylethylketone, acetone, and even smaller amounts of ethanol.[2] In addition, OH contains a catalyst indicated to be dibutyltindilaurate, and, in fact, gels analyzed by SEM-EDS contain small amounts of tin.[3] Today Wacker markets a new version known as OH100, and similar products are sold in the United States under the names *Conservare* OH100 and H100.[4]

From the beginning, OH contained the ketone solvents acetone and methylethylketone (MEK). The alkoxysilane constitutes about 75 percent by volume and the solvents 25 percent, with more MEK than acetone (Bosch, Roth, and Gogolok 1976); the addition of these solvents reduces the viscosity.[5] MEK and acetone also have higher vapor pressures than any of the TEOS components and therefore evaporate preferentially.[6] The OH100 and H100 versions contain no added solvents.

How does the catalyst, dibutyltindilaurate, function in these formulations? It has a relatively long history of use in the silicone rubber industry (Noll 1968). RTV (i.e., room temperature vulcanizing) silicone rubber premixtures typically consist of a viscous but pourable, hydroxyl-terminated silicone resin (1) and TEOS, the cross-linking agent (2). When reacted together, they form silicone rubber(3):

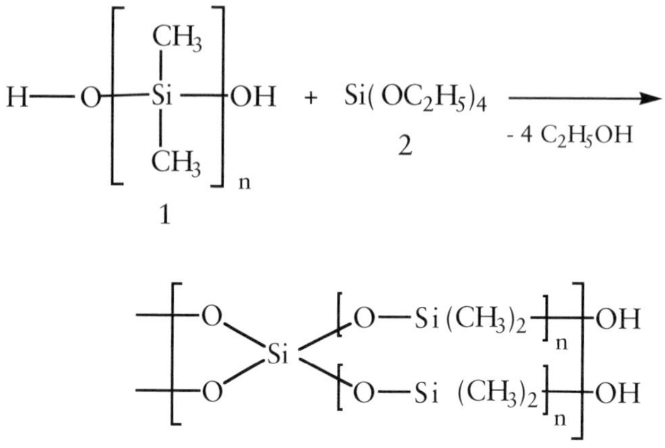

The cross-linking reaction is initiated and accelerated by dibutyltindilaurate in the following manner (Weij 1980) where $R' = (CH_2)_{10}CH_3$:

$$(1)\ R_2Sn(OOCR')_2 + H_2O => R_2Sn(OOCR')OH + R'COOH$$
$$\qquad\quad a \qquad\qquad\qquad\qquad b \qquad\qquad\qquad c$$

$$(2)\ R_2Sn(OOCR')OH + Si(OCH_2CH_3)_4 =>$$
$$\qquad\qquad b \qquad\qquad\qquad\quad d$$

$$R_2Sn(OOCR')\text{-}O\text{-}Si(OCH_2CH_3)_3 + CH_3CH_2OH$$
$$\qquad\qquad e \qquad\qquad\qquad\qquad\qquad f$$

$$(3)\ R_2Sn(OOCR')O\text{-}Si(OCH_2CH_3)_3 + HO\text{-}Si(OCH_2CH_3)_4 =>$$
$$\qquad\qquad e \qquad\qquad\qquad\qquad\qquad\qquad g$$

$$(CH_3CH_2O)_3Si\text{-}O\text{-}Si(OCH_2CH_3)_3\ +\ R_2Sn(OOCR')OH$$
$$\qquad\qquad h \qquad\qquad\qquad\qquad\qquad\qquad i$$

Step (1) is a hydrolysis reaction in which the organotin compound (a) reacts with water to form a tin-hydroxy compound (b) and the organic acid (c). Step (2) is a condensation reaction between the tin-hydroxyl compound (b) and TEOS (d) that creates a tin-siloxane (e) and ethanol (f).

The tin-siloxane (e) is reactive to any silanols (g) and promotes the condensation in Step (3) between the siloxyl part of (e) and the silanol (g) that produces the compound of interest, the TEOS dimer (h). Note that the byproduct (i) of reaction (3)—$R_2Sn(OOCR')OH$—is the same as the primary product (b) of reaction (1). This tin-hydroxy compound is recycled back into the reaction mixture and continues to generate the siloxane bonds that lead to gelation. Therefore, strictly speaking, the hydrolysis product (b)—$R_2Sn(OOCR')OH$—is the catalyst of the reaction and not the original organotin compound (a).[7] This scheme is neatly diagrammed by Weij (1980):

$$\rightarrow Sn-O\overset{\overset{\textstyle O}{\|}}{C}CH_3$$

$$H_2O$$

$$CH_3\overset{\overset{\textstyle O}{\|}}{C}OH$$

$$-Si-OR \qquad \rightarrow Sn-OH \qquad \rightarrow Si-O-Si-$$

$$ROH \qquad -Sn-O-Si- \qquad -Si-OH$$

Small amounts of silanol (Si-OH) must be present to complete the siloxane-forming reactions, and some silanols have indeed been detected by GC-MS in Wacker OH formulations.[8] In addition, Weij (1980:2544–45) makes a passing observation that has importance for the way the dibutyltindilaurate functions as a catalyst in OH: "Evidently partial substitution of the alkoxy groups of the cross-linking agent [his cross-linking agent is TEOS] by siloxy groups leads to products which are much more reactive towards organotin hydroxides [$R_2Sn(OOCR'-OH)$] . . . than the original compounds." This is another way of saying that more condensed silicon atoms undergo condensation reactions more readily than less condensed species such as monomers or dimers; that is, dibutyltindilaurate, though neutral, acts like a base catalyst (see chap. 2). Using oligomers of TEOS rather than monomers helps the catalyst to do its work.[9]

We can speculate further on how the catalyst influences properties of the gel. The two reactive laurate groups are removed during hydrolysis to create lauric acid and dibutyltindihydroxide:

(1) $\{C_4H_9\}_2Sn\{OOC(CH_2)_{10}CH_3\}_2 + 2H_2O =>$

$2CH_3(CH_2)_{10}COOH + \{C_4H_9\}_2Sn(OH)_2$

lauric acid   dibutyltindihydroxide

The lauric acid remains in the gel due to its low vapor pressure and high viscosity.[10] It may also contribute to the reduction of fracturing in the gel by serving as a lubricant that reduces tensile stresses in the pores (Scherer and Wheeler 1997).[11]

In practice, the catalyst in OH is activated by atmospheric relative humidity and water vapor in the stone. Depending on relative humidity and temperature, gel times vary between twelve and twenty-four hours. Although accurately named OH, that is, without a hydrophobing agent,

Figure 4.1

Wacker OH does not mix with liquid water. Note the globule of water suspended in the liquid OH.

the liquid form of the consolidant is immiscible with liquid water due to the alkoxy groups (fig. 4.1), and the gel remains hydrophobic for months after application to stone. This water repellency diminishes as (1) residual alkoxysilanes in the gel's pore liquid hydrolyze and condense, (2) alkoxy groups on the gel itself are hydrolyzed, and (3) the remnants of MEK depart.[12]

One fundamental difference between Wacker OH and Wacker H is that the gel from Wacker H remains permanently hydrophobic, hence the name Hydrophobie.[13] This is the result of including equal amounts of methyltriethoxysilane (MTEOS) and TEOS and their oligomers. H contains the same solvents and catalyst as OH.[14] Samples removed immediately from containers for GC-MS analysis show no evidence that MTEOS components have condensed with TEOS components, which suggests that the two sets of oligomers are formed separately and blended together to create the consolidant. With the exception of water repellency, the properties of the gel formed from H appear to be similar to those of OH, although gels formed in glass Petri dishes from H are less fractured than are gels of OH (fig. 4.2). The lower functionality of MTEOS—three sites for condensation versus four in TEOS—makes these gels more compliant during drying.

As mentioned earlier, the main products that Wacker Chemie markets today are OH100 and H100, both of which do not contain added solvents.[15] For OH100 the viscosity is nearly twice that of the ear-

Figure 4.2

Gels derived from Wacker OH on the left are fractured (Wacker OH containing 75% TEOS monomer, dimer, and oligomers, 25% of the solvents methylethylketone and acetone, and 1% of the catalyst dibutyltindilaurate). Gels derived from Wacker H on the right (with 37.5% TEOS monomer, dimer, and oligomers and 37.5% MTEOS monomer, dimer, and oligomers) are less fractured due to the lower functionality of MTEOS versus TEOS that makes for more compliant gels during drying.

lier version, but the flash point increases from 2°C to 40°C, making it considerably safer to use (Wacker-Chemie 1999). Although this increase in viscosity may seem large, the actual value is still low and is not likely to significantly affect the performance of the consolidant. *Conservare* products marketed by ProSoCo in the United States went through the same transformations to OH100 and H100. Some batches of the earlier version of *Conservare* OH analyzed over the past several years were found to also contain mineral spirits in addition to MEK and acetone.[16]

During the past three decades, other companies have produced and marketed stone consolidants similar to Wacker OH and H. For example, Keim's Silex OH and H are also based on oligomeric TEOS and MTEOS. Like the Wacker products, early versions contained mostly monomer and some dimer, while current versions are oligomeric; the solvent is toluene and the catalyst dibutyltindilaurate.[17] Gels formed in glass Petri dishes are less fractured than the Wacker or *Conservare* equivalents (fig. 4.3), probably due to the lower vapor pressure of toluene versus the mixture of acetone and MEK for Wacker.[18]

The T. Goldschmidt products Tegovakon V and Tegovakon T have competed for market share of stone consolidants from the beginning.[19] The consolidating ingredients are the same as those in Wacker, *Conservare*, and Keim products—oligomeric TEOS and MTEOS—but are weighted more toward monomers and dimers.[20] Tegovakon V is the OH equivalent, and Tegovakon T is the hydrophobic version similar to H. The catalyst is also a tin compound. Solvents compose 34% of the formulation: 17% naphtha (a mixture of low-boiling, mostly aromatic hydrocarbons) and 17% ethanol.[21] The gels from both Tegovakon V and Tegovakon T are much less fractured than other formulations based on catalyzed TEOS or MTEOS, and Tegovakon T is even less fractured than V (see fig. 4.4).

Remmers, located in Löningen, Germany,[22] has marketed a product line under the name Funcosil, which until recently has consisted of four main products: Funcosil OH, Funcosil 100, Funcosil 300, and Funcosil 510.[23] Funcosil OH appears to be the equivalent of Wacker OH: it contains 75% (by weight) of oligomeric TEOS, and the remainder is methylethylketone and dibutyltindilaurate. Sattler and Schuh (1995) indicate that Funcosil OH deposits about 30% w/w of gel. Funcosil 300 is another catalyzed formulation that deposits 30% w/w of gel but is solvent-free and therefore similar to Wacker OH100. To achieve

Figure 4.3

Gels derived from Keim OH (left) and Keim H (right) are less fractured than their Wacker equivalents. This is probably due to the lower vapor pressure of toluene as compared to the MEK-acetone mixture found in Wacker H.

Figure 4.4

The Tegovakon V gel on the left is less fractured because the naphtha-ethanol solvent mixture is low in vapor pressure. It is the least fractured of all catalyzed TEOS and TEOS-MTEOS formulations. The Tegovakon T gel on the right is even less fractured than those from Tegovakon V because of the lower functionality of MTEOS.

the same mass return of gel with no solvent, the oligomeric TEOS component must be slightly less condensed than in Funcosil OH.[24] Funcosil 100 and Funcosil 510 are also catalyzed, contain additional solvent, and deposit 10% to 45% w/w of gel respectively.

Another German company, Interacryl of Frankfurt, sells four ethyl silicate stone strengtheners: Motema 20, 28, 29, and 30. The numbers represent the approximate w/w% of SiO_2 in the formulation. The solvent is ethanol, and, as such, these are the only TEOS products that have the possibility of forming a solution with liquid water in the stone. Motema 28 and 30 are uncatalyzed; Motema 20 and 29 contain < 0.1% w/w and < 0.3% w/w catalyst, respectively. Although the catalyst is not specifically identified, it is probably dibutyltindilaurate because gels from Motema 20 and 29 contain small amounts of tin.[25] The Brookfield viscosities at 23°C for the group of products are < 1, 4, 5, and 6 mPa•sec.[26] The highest value is in the range where concern arises for depth of penetration and penetration into smaller pores.

Product information sheets and publications by Koblischek (see, e.g., Koblischek 1995) of Interacryl emphasize the importance of using –prehydrolyzed (more accurately, precondensed) silicic acid ethyl ester for better results in consolidating stone and for user safety. Of course, prehydrolyzed silicic acid ethyl ester is another way to describe oligomeric TEOS. Koblischek states that using uncatalyzed, oligomeric TEOS results in a smoother profile of strength versus depth. Despite the undoubtedly longer gel times, he prefers these uncatalyzed formulations. When catalysts are present in the Motema formulations, their concentrations are lower than other ethyl silicate consolidants: < 0.1% and < 0.3%, as compared to about 1%.

Rhone-Poulenc (now Rhodia)[27] has also formulated stone consolidants since the early 1970s.[28] Their work led to three products, RC70, RC80, and RC90.[29] RC70 contains 70% oligomeric TEOS, with the balance made up of white spirits (a mostly nonaromatic hydrocarbon solvent)[30] and a catalyst. According to Rhone-Poulenc's older product literature, the catalyst is "an ethyltinsiloxane . . . or . . . an organoditinsiloxane of the general structure":[31]

$$R' \quad R'$$
$$RO\text{-}Sn\text{-}O\text{-}Sn\text{-}O\text{-}Si\text{-}(OR)_3$$
$$R' \quad R'$$

This catalyst would be expected to react in a manner similar to dibutyltindilaurate. Returning to the reaction sequence for dibutyltindilaurate shown previously, it is evident that a stanno-siloxane compound similar to the catalyst in RC70 is an *intermediate* in the reaction of dibutyltindilaurate with TEOS. Apparently, RC products have used this intermediate form as their catalyst.

Both RC80 and RC90 contain water-repellent components in addition to oligomeric TEOS, white spirits, and the tin-siloxane catalyst. In RC80 this water-repellent component is reported to be a methylsilicone resin; the oligomeric TEOS content is slightly reduced, and 8% of the white spirits is replaced with xylenes, presumably to help to dissolve the methylsilicone resin. For RC90 the water-repellent component is methyl*phenyl*silicone, and the 8% xylenes are replaced by 6% toluene; the other components remain the same. Methylphenylsilicone imparts greater water repellency than methylsilicone due to the large, all-hydrocarbon phenyl group.

Adding these resins reduces the TEOS content, and the resins themselves have a low functionality: in fact, they may only have reactive end groups. In general, the gel would be less condensed and/or the TEOS-derived part of the gel would contain segments of silicone resin. In the gel-forming process fewer cracks appear because the gel is more compliant. The large phenyl groups in RC90 enhance the compliancy or elasticity of the gel—a property used by Schmidt and Fuchs (1991) to create flexible, protective coatings for medieval glass. RC90 gels formed in Petri dishes are somewhat less fractured than most TEOS-derived gels (fig. 4.5). RC80 gels turn white at gelation, perhaps because the methylsilicone resin forms a separate, insoluble phase in the TEOS matrix.

A Japanese company, COLCOAT, founded in 1956, manufactures a series of ethyl silicate products, one specifically for the consolidation of porous materials (SS-101).[32] Nishiura (1987a, 1995) claims that the solvent is a mixture of MEK and acetone and that a tin catalyst is employed in the formulation.

The components of the commercial formulations discussed so far are premixed, in a single container, and ready for use.[33] Brethane, the product developed at the Building Research Establishment in the 1970s by Leslie Arnold and Clifford Price under the supervision of David

Figure 4.5

The gel derived from RC90 on the right is somewhat less fractured than most ethyl silicate–derived gels due to the inclusion of methylphenylsilicone. Given the large size of the phenyl group, it is surprising that the gel does not show even less fracturing. The RC80 gel also exhibits less fracturing. The whiteness may result from the precipitation of the methylsilicone resin from the TEOS gel.

Honeyborne, is different in two important ways: it is a four-component system packaged in three containers whose contents are mixed at the time of application; and the consolidating ingredient is methyltrimethoxysilane (MTMOS) monomer.[34] The other ingredients are water, industrial methylated spirits (denatured alcohol), and a catalyst called Manosec Lead 36. Manosec Lead 36 is a 36% solution of lead naphthenate in a hydrocarbon solvent.[35] The gel from Brethane is much less fractured than are gels from other commercial products (fig. 4.6), due primarily to the trifunctionality of MTMOS.[36] The added water, 13 ml for every 64 ml MTMOS, is the stoichiometric amount for complete hydrolysis and condensation (1.5 moles of water to 1 mole of MTMOS).[37] Ethanol is the co-solvent that makes the water, MTMOS, and catalyst miscible. The lead naphthenate catalyst is diluted in dried white spirits, 6% weight/volume Manosec Lead 36/white spirits, and 1% by weight of this solution is used to catalyze the reaction.[38] As a coordination compound of a metal and a large organic acid, this catalyst probably functions in Brethane in a manner similar to dibutyltindilaurate in OH:[39] the catalyst is hydrolyzed by water; the hydrolyzed catalyst reacts with MTMOS to produce a lead-siloxane intermediate; and the activated, lead-siloxane intermediate reacts rapidly with other MTMOS molecules or more condensed MTMOS oligomers and creates the siloxane bonds leading to gelation. Because liquid water is added directly to the mixture, gel times for Brethane are usually shorter (2–6 hours) than for other alkoxysilanes systems (more than 12 hours).[40] Due to the denatured alcohol in the formulation, Brethane can also form a homogeneous solution with water that may be present in the pores of the stone.

Noncommercial Formulations

The most common alkoxysilane-based noncommercial formulation used for the consolidation of stone can hardly be called a formulation: it consists only of monomeric methyltrimethoxysilane or neat MTMOS.[41] Introduced by Hempel and Moncrieff in the late 1960s, neat MTMOS

Figure 4.6

Gel derived from Brethane. The Brethane gel is among the least fractured because the alkoxysilane component consists entirely of the trifunctional MTMOS.

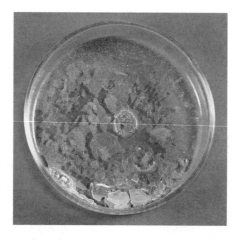

and other MTMOS-based consolidants have been used primarily in England and to a lesser degree in the United States.[42] In this form MTMOS reacts with water vapor to initiate the hydrolysis and condensation reactions that ultimately lead to gelation. The water vapor derives from the environment that surrounds the stone during treatment and from liquid water and water vapor in the stone.[43] The reactions are slow under these conditions, and, as Arnold and Price (1976) pointed out, the curve of viscosity versus time does not conform to the classic shape of rapid increase at the point of gelation noted in chapter 2. Gelation may take weeks, and, as Charola, Wheeler, and Freund (1984) have shown, there is a nearly direct relationship between gelation time and relative humidity. Higher relative humidities (greater than 50%) also produce more gel with more fractures (fig. 4.7).[44]

Hempel and Moncrieff, either intuitively or explicitly, recognized that liquid water inside a stone represents a problem for consolidants such as neat MTMOS that, by themselves, are unable to form homogeneous solutions with this water. When pore-invading MTMOS meets resident liquid water, a heterogeneous reaction occurs. The reaction produces quite a different gel from that produced from water vapor alone.[45] To overcome this problem, Hempel pretreated the stone with a co-solvent for water and MTMOS—either alcohols or cellosolves.[46] These solvents initially formed a solution with water in the stone and later allowed the smooth incorporation of MTMOS. In addition to the water immiscibility of neat MTMOS, they addressed its slow reaction rate by adding liquid water and co-solvents.[47] Starting with alcohols (methanol, ethanol, and 2-propanol) and later ethyl cellosolve, these solvents created homogeneous reactions in the stone, less fractured gels, and shorter gel times.[48] Hempel and Moncrieff addressed the need to speed up the reaction of MTMOS by adding what at the time would have been a rather unique catalyst, titanium isopropoxide. Like alkoxysilanes (or silicon alkoxides), titanium alkoxides are gel forming but react much more quickly than their silicon analogues.[49]

Noncommercial formulations based on TEOS are also reported in the literature. Laurie (1926a) used HCl-catalyzed ethyl silicate as a treatment for limestone, as did Lewin (1972), who also added a small amount of water. Chiari (1980) used a similar HCl-catalyzed oligomeric mixture of TEOS on adobe, and there are references to the use of uncatalyzed ethyl silicate 40 on adobe.[50]

Figure 4.7

Neat methyltrimethoxysilane can produce gels by reacting with atmospheric moisture. The gel time, amount of gel, and degree of fracturing are all dependent on the relative humidity. From left to right, gels formed at 25%, 50%, and 75% RH.

It might seem contraindicative to use acid catalysts on acid-soluble substrates such as limestone, and, in fact, Wheeler demonstrated that HCl-catalyzed solutions of water, TEOS, and ethanol made Monks Park limestone *weaker* than untreated stone.[51] Goins, Wheeler, and Fleming (1995), following Vega and Scherer (1989) prepared HCl-catalyzed solutions and removed the HCl and excess ethanol by roto-evaporation. This solution provided good strength increases for Monks Park limestone but no better than increases obtained from *Conservare* OH. The main purpose of the work of Goins and colleagues was to examine the role different solvents played in the performance of TEOS-based consolidants. Solvent type did influence the performance (Domaslowski 1987–88; Fowkes 1987), but it was also seen that solutions more concentrated in TEOS gave higher strength increases.

As several researchers have pointed out, higher strength increases are not necessarily better. The main concern is overconsolidation near the surface with an abrupt change in strength at some depth. This is the same concern expressed by Koblischek (1995). Ettl and Schuh (1989) produced a smooth transition between weathered material on the exterior and the sound, unweathered stone substrate by consolidation with Wacker OH diluted with additional solvents.[52]

Noncommercial formulations have seen very little use since the heyday of MTMOS, from about 1968 to 1985.[53] Most commercial formulations[54] are based on oligomeric TEOS, and those that are not are a mixture of TEOS and MTEOS, methylsilicone resin or methylphenylsilicone resin. Commercial products may be solvent-free or contain solvents with a range of vapor pressures, viscosities, water solubilities, and toxicities. Many formulations are catalyzed with "neutral" tin compounds. This wide array of compounds can be made into many different products, all of which come to the user in a single container, ready to use. Brethane is exceptional on almost all counts: its consolidating compound is MTMOS, not TEOS or MTEOS; its catalyst is the more toxic lead naphthenate, not tin-based; and its components, in three containers, must be mixed at the time of use.

There are compelling rationales for the different components in the formulations described in this chapter: water solubility of solvent, less fractured gels, smooth strength-depth profile, and so on. It is surprising to conclude after examining much of the literature, however, that there does not appear to be a discernible systematic difference in performance among the many commercial products. This does not mean that such a difference does not exist, only that it is not discernible in the literature. More is said on this subject in chapters 5 and 6.

Notes

1. Patent applications attempt to cover a wide range of potential products. The U.S. Patent for the Wacker products is quite typical in this regard.

2. The ethanol is not an added ingredient but the result of some hydrolysis of TEOS in the container. All spectra were collected on the Hewlett-Packard Model

5992 Gas Chromatograph–Mass Spectrometer using a capillary column with a stationary phase of OV-101, and helium as the mobile phase. The programming remained largely the same for all runs:

1) starting temperature = 40°C and held for 1 min.
2) temperature increase rate = 16°C per min.
3) final temperature = 220°C
4) injection port temperature = 250°C
5) split mode
6) electron beam of 70ev
7) electron multiplier setting = 1400–2600, varied as sensitivity of instrument changed
8) solvent elution time = 1.0 min., sometimes longer to allow for the elution of high concentration components
9) flow rate = 1 ml per min.
10) samples per 0.1 amu = 2
11) mass peak detection threshold = 4000
12) injection volume = 1 microlitre, 10 microlitre maximum volume Hamilton syringe
13) sample concentration = 0.1% v/v in diethylether–neat reaction mixture.

3. SEM-EDS stands for scanning electron microscopy–energy dispersive spectroscopy. It is a technique for performing elemental analysis. These analyses were performed by Mark Wypyski, Research Scientist at the Sherman Fairchild Center for Objects Conservation of the Metropolitan Museum of Art.

4. ProSoCo, Inc., Lawrence, Kansas, manufactures, packages, and sells a full range of products for building conservation.

5. Acetone has a viscosity of 0.3029 and methylethylketone 0.378 cSt at 25°C, that is, much lower than TEOS monomer (0.7180) or any of its oligomers. The addition of these solvents reduces the overall viscosity of the formulations to about 1.5 cSt. The vapor pressure of acetone is 231 mm Hg at 25°C, 90.6 mm Hg for methylethylketone, and 5 mm Hg for TEOS (the oligomers would have even lower vapor pressures). Data are from Riddick, Bunger, and Sakano 1986 and from contract research by the author supported by the Getty Conservation Institute.

6. Raoult's Law shows that MEK and acetone will carry some of the TEOS components with them as they evaporate. In fact, ProSoCo suggests a solvent wash at the end of treatments with *Conservare* OH to prevent surface gloss. This solvent wash removes excess OH from the surface of the stone by evaporation.

7. A catalyst is a compound that increases the rate of reaction. However, it is not consumed but is regenerated to continually perform its function throughout the reaction period.

8. Analyses were performed with a Hewlett-Packard 5890 Gas Chromatograph and a 5970B MSD with acquisition parameters similar to those described in note 2.

9. Oligomers are low-molecular-weight polymers and more readily react with base catalysts to produce siloxane linkages.

10. Data are not available for lauric acid. However, the vapor pressure of the smaller nonanoic acid is 0.0014 mm Hg at 25°C. Similarly, the viscosity of nonanoic acid is 7.15 cSt at 25°C and lauric acid's would be even higher. Data are from Riddick, Bunger, and Sakano 1986. The viscosity of palm kernel oil which is about 50% lauric acid is 31 cSt at 100°C. Therefore, this acid is not very mobile. Data from Swern 1979.

11. Brus and Kotlik (1996) made a study of cracking in Tegovakon and other gels. They concluded that the catalyst promoted cracking.

12. MEK is only 24% soluble in water at 20°C (Riddick, Bunger, and Sakano 1986).

13. Eventually, even the water repellency of H will diminish or disappear as the Si-C bonds of the methyl groups are cleaved.

14. Analyses were performed with a Hewlett-Packard 5890 Gas Chromatograph and a 5970B MSD with acquisition parameters similar to those described in note 2.

15. This formulation is more in compliance with volatile organic components (VOC) laws in the United States and other parts of the world. The difference between OH and OH100 in the amount of VOCs is small. Perhaps of equal importance is that OH100 allows the user to dilute the product with a solvent or solvents of choice.

16. Mineral spirits can be quite low in viscosity so that their inclusion in *Conservare* OH will have only a small effect on viscosity. The flash point of mineral spirits is, however, quite low.

17. This analysis was performed with a Hewlett-Packard 5890 Gas Chromatograph and a 5970B MSD with acquisition parameters similar to those described in note 2.

18. The vapor pressure of toluene is 28 mPa at 25°C, much lower than acetone, MEK, or mineral spirits. This lower vapor pressure reduces the capillary tensile stress in the pores.

19. Munnikendam (1967) was one of the first to mention the Goldschmidt products in the conservation literature.

20. Analyses were performed with a Hewlett-Packard 5890 Gas Chromatograph and a 5970B MSD with acquisition parameters similar to those described in note 2.

21. Naphtha can have a wide range of vapor pressures depending on which fraction of the petroleum distillate is taken. It is not water-soluble and is more toxic than many other solvents due to its aromatic components.

22. Remmers was established in 1949. The main offices are in Löningen and Munster, and a satellite office operates in Burgess Hill, West Sussex, England.

23. Funcosil 100 is a more dilute form of OH.

24. For every 100 grams of Funcosil OH, only 75 grams are TEOS oligomers that must produce 30 grams of SiO_2. If those same oligomers made up 100% of the solution (i.e., no solvents), then 100 grams of this solution would deposit more than 30 grams of silica gel. Therefore, the oligomeric component of Funcosil 300 must be less condensed than the oligomeric component in Funcosil OH.

25. SEM-EDS analyses were performed by Mark Wypyski, Research Scientist at the Sherman Fairchild Center for Objects Conservation of the Metropolitan Museum of Art.

26. For determining Brookfield viscosity, see ASTM D789 and D4878.

27. This name change occurred sometime in the late 1990s.

28. The earliest reference to Rhone-Poulenc's involvement in stone conservation is found in the work of Hempel and Moncrieff (1972). They experimented with a form of MTMOS from Rhone-Poulenc called X54-802.

29. This information comes from the Material Safety Data Sheets (MSDS) for RHOXIMAT RD RC70, RC80, and RC90, Rhodia, 9 April 1997.

30. Like naphtha, white spirits can have a wide range of vapor pressures depending on the petroleum distillate it represents. Being mostly nonaromatic, it is much safer to breathe than naphtha.

31. This information comes from the product literature of Rhone-Poulenc Italia, October 1988.

32. This information was taken from the COLCOAT Co., Ltd., Web site on 14 December 2000.

33. The shelf life of most of these single-container systems is given at about 12 months. It can be extended if the environment is very dry. *Conservare* OH has been stored for several years when the container is placed in a sealed plastic bag with "dry" silica gel. Since the catalyst is not mixed with the alkoxysilane component until the time of use, the shelf life of Brethane could be much longer. However, Butlin, Yates, and Martin (1995) indicated that the catalyst may not be stable.

34. The name Brethane is apparently a contraction of BRE (Building Research Establishment) and methane or ethane. It is no longer commercially available but can easily be prepared from the components listed in the text.

35. This information comes from the Manchem Ltd., Manchester, England, product literature, *Driers and Drying* (1981).

36. It is not known why Brethane gels are often white.

37. Hydrolysis reactions consume water, and condensation reactions produce water. Therefore, theoretically, only 1.5 moles are necessary for complete hydrolysis. However, it is clear from subsequent work in the field of sol-gel science that molar ratios upwards of 8:1 H_2O:alkoxysilane with acid catalysis are necessary for complete hydrolysis. See Brinker and Scherer 1990.

38. This catalyst was typically used as a paint drier (*Driers and Drying* 1981).

39. The naphthenic acid byproduct of this hydrolysis is "a group of saturated higher fatty acids derived from the gas-oil fraction of petroleum by extraction with caustic soda solution and subsequent acidification" (*Condensed Chemical Dictionary* 1981).

40. Again, Butlin, Yates, and Martin (1995) suggest that gel times may not be reliable because of the instability of the catalyst.

41. Despite some limitations, neat MTMOS was shown in chapter 3 to give good strength increases to some sandstones.

42. The references list many works that support this statement. In England: Hempel and Moncrieff 1972; Larson 1983; Bradley 1984; Hanna 1985; Dinsmore 1987; Miller 1992. In the United States: Charola 1983; Wheeler 1984; Grissom 1999. MTMOS was also tested in Egypt by Saleh (1992), and Hempel used it in Italy.

43. Koblischek (1995) also refers to the Helmholtz stratum of water molecules on solid surfaces that contributes to the water necessary for the reaction of alkoxysilane formulations that do not have water added to them.

44. It is surprising that so much fracturing occurs above 50% RH, given that the gel times are so long.

45. The gel that forms from this heterogeneous reaction is dense and white. It can clog pores, preventing further penetration of consolidant, and creates a front through which neither water vapor nor liquid water can pass.

46. Like ethanol, ethyl cellosolve or 2-ethoxyethanol serves as a co-solvent and has a low vapor pressure of 5.3 mm Hg at 25°C, much lower than ethanol at 58 mm Hg (Riddick, Bunger, and Sakano 1986).

47. The resulting formulation is similar to Brethane but without the catalyst. The work on Brethane had probably not yet begun at this time.

48. The effect of water and solvent content and solvent type on gel time and relative fracturing of gels was studied much later by Wheeler in contract research for the Getty Conservation Institute. See Wheeler et al. 1992.

49. The idea of using titanium isopropoxide probably comes from Noll (1968). After Hempel, it was picked up by Larson (1983), and it was also discussed by Bradley (1985) and Miller (1992). Wheeler had limited success with mixtures of TEOS and titanium isopropoxide as stone consolidants. This unpublished work was performed under a research contract for the Getty Conservation Institute.

50. In unpublished field reports, C. Selwitz refers to the use of ethyl silicate (Silbond 40) on adobe at Fort Selden, New Mexico.

51. From unpublished contract research for the Getty Conservation Institute.

52. Laurie (1926a, 1926b) already suggested using a diluted form of ethyl silicate, but this suggestion goes against the current trend of reducing volatile organic components (i.e., solvents) in consolidants.

53. They have been used to some degree on adobe.

54. Other commercial products are mentioned in the literature or are used by practitioners. In Italy, Alaimo et al. (1996) mentions Rankover as a source of ethyl silicates. CTS, also in Italy, markets an ethyl silicate formulation known as Estel 1000, a water-repellent version based on ethyl silicate and a "siloxane" (Estel 1100) and a biocidal version also based on ethyl silicate (Bio Estel). Finally, Phase, a company in Bologna, sells an ethyl silicate–based product "without acid catalyst."

Chapter 5
Practice

Conservators ultimately want to know how, when, and under what conditions to use alkoxysilanes. That kind of treatment map is not easily drawn from the incomplete and conflicting information that is available. Nonetheless, guidelines can be derived from common sense, inference from laboratory studies, and especially from practitioners' experiences. This chapter discusses pretreatment conditions and activities, application techniques, consumption of the consolidant, application schedules, and conflicting activities.

The initial and basic guidance presented here comes from the product literature. For most alkoxysilane consolidants this literature indicates that air temperature and the surface temperature of the stone should be between 10°C and 32°C.[1] Dark-colored stones are more susceptible to surface heating by insolation; as surface temperatures can reach 80°C (Winkler 1994), protection from direct sunlight with tarpaulins or awnings is indicated. These consolidants should not be applied to wet surfaces or when there is a chance of rain (see *Moisture and Wetness* below). Relative humidity should be between 40% and 80% (Charola, Wheeler, and Freund 1984). Environmental temperature and relative humidity can be measured with many devices, and stone temperatures can be determined with surface thermometers or remote handheld thermometers (fig. 5.1).

Figure 5.1
The device at left is attached to the stone with silicone grease that must be cleared after the thermometer is removed. The infrared thermometer on the right is used remotely.

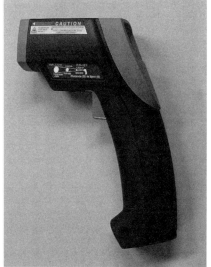

Preconsolidation

When surface conditions such as scaling, flaking, exfoliation, or severe granular disintegration exist (fig. 5.2),[2] an immediate and full-scale application of alkoxysilane consolidants can sometimes cause dislodgement and loss of stone fragments or grains.[3] In these cases an initial stabilization is carried out—a process often referred to as preconsolidation. This stabilization is performed to allow cleaning or removal of salts (Lukaszewicz 1996b) or simply to hold fragments in place during a complete, in-depth consolidation with alkoxysilanes.

Typically, scales and flakes are secured with adhesives by local injection, often supplemented by the application of facing tissues (fig. 5.3) (Nonfarmale 1976). The choice of adhesive may depend on the type of alkoxysilane consolidant that will be employed subsequent to preconsolidation. MTMOS readily dissolves most acrylic resins and polyvinyl acetates (PVAs). As Dinsmore (1987) points out, acrylics such as ACRYLOID B72 and ACRYLOID B48N[4] used to set down scales or flakes may need to be renewed after application of uncatalyzed MTMOS as the in-depth consolidant.[5] She found that polyvinyl alcohol (PVOH) used for setting down tissue was unaffected by the second-stage consolidation with MTMOS.[6] In contrast, Bradley (1985) found that it is harder to remove the tissue with catalyzed formulations such as Wacker OH. For painted stone to be treated with MTMOS, Mangum (1986) found preconsolidation with gelatin superior to carboxymethylcellulose, ethylcellulose, ACRYLOID B72, and PVA for stabilization of the paint layer before consolidation with MTMOS.

A broader range of resins can be used as preconsolidants in conjunction with both OH- and H-type consolidants because acrylics, PVAs, PVOHs, and cellulose resins are insoluble or sparingly soluble in OH and H.[7] These resins are even less soluble (essentially insoluble) in OH100 formulations, that is, those that do not contain the solvents acetone and

Figure 5.2

Forms of deterioration such as severe granular disintegration, scaling, flaking, and exfoliation pictured here should be stabilized or adhered before full-scale consolidation with alkoxysilanes (Fitzner et al. 1995). Photos: Berndt Fitzner

Figure 5.3

Stabilization of some forms of deterioration may require facing tissue to prevent
losses during treatment with alkoxysilanes
(Ottorino Nonfarmale 1976).

methylethylketone.[8] Thus Grissom (1996) used methylcellulose as a
preconsolidant for lime-plaster sculptures later consolidated with
Conservare OH, and many practitioners have used ACRYLOID B72
before application of OH.[9] In addition, OH-type consolidants have themselves been used as preconsolidants before cleaning. Lukaszewicz (1996)
carefully applied Funcosil OH[10] before cleaning and full consolidation,
Zanardi et al. (1992) applied Wacker OH as a preconsolidant followed
by water washing and additional Wacker OH, and Stepien (1988) used
undiluted Wacker OH in a similar exercise for sandstone. For limestone,
Stepien performed preconsolidation with limewater, followed by cleaning
and full-scale consolidation with Wacker OH. Verges-Belmin (1991) used
Wacker OH before cleaning sulfated marble, followed by RC90 because
it bridged larger gaps (60–400μm) than OH (less than 50 μm). More
recently, solutions of ammonium hydrogen tartrate in a product known
as HCT has been used as a preconsolidant for sugaring marble and, to a
lesser degree, for limestone and calcareous sandstones.[11]

Moisture and Wetness

Another important consideration before treatment is wetness—both
noticeable wetness on the stone and excessive moisture in the stone.
Wetness may be due to direct rainfall, elements on buildings and forms
on sculptures that channel and hold water, poorly maintained water
management systems, and broad or localized environments with consistently high relative humidities.

Even if stone is not visibly wet it may contain moisture from salts.
Salts become hygroscopic, this is, attract moisture and eventually deli-
quesce, if the relative humidity of the surrounding environment exceeds

the equilibrium relative humidity of the salt (Price and Brimblecombe 1994). The resulting wetness must be removed before treatment can commence. Finally, biological growth, in addition to being a direct and indirect source of deterioration, can complicate treatment by retaining or trapping moisture.

Avoiding wetness from rainfall can be minimized by consulting weather reports, scheduling work when rain is less common, and protecting the areas to be treated with plastic, framed temporary shelters, or covered scaffolds. But even with the best planning, the stone may be wet at the time of treatment. Several strategies have been adopted for eliminating or coping with wetness. For small objects or small areas of wetness, the warm, dry air from a hair dryer or heat gun suffices. The stone should not be overheated, particularly large-grained marbles, in which the risk of damage is significant (Sage 1988). Hempel's "solvent" techniques described in chapter 4 can be used for small- to medium-sized objects or areas, perhaps up to about 100 ft^2 or 10 m^2. In one approach, he applied volatile, water-miscible solvents such as acetone or ethanol that carried away most of the water through evaporation.[12] In another approach, he applied water-miscible solvents of lower volatility (ethylcellosolve or ethoxyethanol). While these solvents do not evaporate quickly enough to dry the stone, they allow alkoxysilanes applied afterward to form solutions with water in the stone. On scales larger than 100 ft^2 the solvent approach is both unsafe and impractical. A scale-up of the hair dryer technique entails enclosing the area to be treated—usually with tarpaulins or heavy plastic sheeting—and delivering warm, dry air with space heaters. Because the consolidants are highly flammable, the heaters should be turned off during their application.

It is quite uncommon to encounter stone that is too dry. However, in Egypt, insolation and low humidity often desiccate stone. The dryness stalls the reaction and together with the heat causes much of the alkoxysilane to evaporate before gelling.[13] Kariya[14] addressed these problems in her work at Luxor Temple in the following manner: to provide some moisture to drive the gel-forming reactions, she placed wet cotton compresses on the stone's surface and covered the compresses with plastic sheeting for 30 minutes; to deter the evaporative effects of the sun, consolidation treatment was carried out in canvas shelters; to further limit evaporation the stone was loosely covered in plastic after application of the consolidant.[15]

Biological Growth

In connection with the consolidation of stone, the primary issues with biological growth are that it holds moisture and that it traps moisture when it covers the surface of the stone (which also limits ingress of the consolidant). Biological growth takes many forms on stone substrates: bacteria, fungi, algae, lichens, mosses, vines, creepers, and trees (fig. 5.4). Bacteria and fungi do not interfere with the application of consolidants because they are microscopic in scale.[16] In contrast, algae are visible as

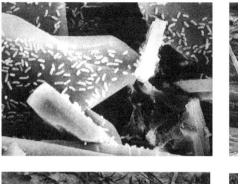

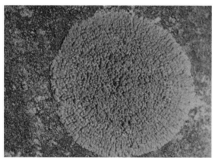

Figure 5.4

Biological growth takes many forms. Microbiological forms such as bacteria and fungi do not generally interfere with consolidation. Algae, lichens, mosses, and higher plants can cause complications during treatment due to water they carry or the inhibition to the drying of the stone. Photos: Robert Koestler. (To view this figure in color, see note on the copyright page.)

green and brown films that keep moisture in the stone. Even more problematic are lichens, which can have large fruiting bodies that cover stone surfaces, often in dramatic fashion. These fruiting bodies hold moisture and can form a nearly complete mat that slows the evaporation of water from inside the stone. Depending on the type of lichen, the fruiting bodies may also be well attached to the stone.[17] Mosses, although often thick and laden with water, are generally not well attached. Finally, vines and creepers can cover and shade surfaces, making them damp or inaccessible for treatments.

The removal of algae, lichens, mosses, and vines before treatment depends on the condition of the stone and their degree of attachment. Algae and mosses can often be removed mechanically. Initial drying of algae allows much of it to be flaked off; residues can be removed by rewetting and gently agitating with brushes. The weakly attached mosses can be picked off and residues brushed away. Many lichens, particularly the forms that penetrate below the surface of the stone, have a much stronger attachment. Direct and complete mechanical removal without some preconditioning risks pulling away loose grains or even larger pieces of stone. Drying lichens as a first step has its risks because they can become mechanically stiffer and therefore harder to remove. In addition, lichen mats shrink with drying and may tear off small pieces of stone. First wetting the lichens with water, solvents, or mixtures of the two makes them more compliant, and removal is achieved with less damage to the stone. Application of a progression of mixtures from water-rich to solvent-rich—using solvents such as acetone and ethanol—produces conditions that allow removal of most of the lichen, after which most of the water is removed by evaporation.[18] It is also desirable to remove dead or dying lichens in a similar manner. In either case there is a risk of staining the stone by wetting dark or deeply colored lichens with ethanol. For vines and creepers, there is little remedy other than complete, mechanical removal to provide the means for drying the stone and access for consolidants.

Another approach for dealing with most biological growth is to apply biocides or biostats. The palette of biocides that are not harmful to people and the environment is limited. In addition to these safety issues, there are concerns about the effect they have on the stone and on the reactions of the alkoxysilanes with which they come in contact.[19] Some biocidal tin compounds shorten the gel time of some formulations, while the effect of quaternary ammonium products such as D2 are not known.[20] Therefore, when biocidal treatment is complete, thorough rinsing and drying must take place before application of the alkoxysilane.

A final note on biological growth: alkoxysilane consolidants have been shown to slow or suppress growth. This may be due to the toxic nature of some of the catalysts such as dibutyltindilaurate in many formulations and lead naphthenate in Brethane,[21] as well as the temporary removal of water by the reacting alkoxysilanes.[22]

Salts

Stone containing salts often requires consolidation because salts are a common and powerful source of deterioration. This deterioration takes the forms of granular disintegration (powdering, sanding, etc.), scaling, flaking, and exfoliation, depending on the general and local concentration of the salts. Already mentioned are salts that have deliquesced and create a moisture or wetness problem that complicates treatment with alkoxysilanes. Stone in this condition should be dried slowly to reduce damage from crystallization.[23] One of the goals of consolidation is to reduce losses during the removal of salts. It was noted early on that neat MTMOS and solutions of water, solvent, and MTMOS cured and consolidated stone in the presence of salts.[24] It was originally thought that the deposited gels would encapsulate and thereby deactivate salts, but this was soon found incorrect (Berry and Price 1989). In fact, salt removal helps to preserve the consolidating gel. Frogner and Sjoberg (1996) and Thickett, Lee, and Bradley (2000) found that salts break down the silicate network created by OH-type formulations.

In the case of neat MTMOS, Bradley and Hanna (1986) showed that treatment actually mobilized salts in Egyptian limestone and that efflorescences formed on the surfaces of the stone during curing of the consolidant. They suggested desalination immediately after curing of the alkoxysilane and indicated that, surprisingly, even with this hydrophobic consolidant, salt removal was not only possible but also more efficient after treatment. Salts were removed with paper pulp and distilled water.[25] Lukaszewicz (1996) reported that salt removal is more difficult in sandstone because of the conformal gel coatings.

In outdoor environments practitioners were able to consolidate salt-laden stone with neat MTMOS and catalyzed ethyl silicate solutions.[26] However, the type of salt that is present could condition the stone to the salt's equilibrium relative humidity (RH_{eq}). Consolidants such as Wacker OH contain catalysts activated by ambient relative humidity, and in this case the time to gelation would be dependent on that RH_{eq} (see also chap. 4). For Brethane, the lead naphthanate catalyst is reported to be destabilized in the presence of salts, making gel times unpredictable (Arnold 1978; Butlin, Yates, and Martin 1995).

Application Techniques

Several techniques (and devices) for applying alkoxysilanes have been used over the past several decades (fig. 5.5). For large surfaces such as buildings, monuments, and sculptures, pump sprayers are most common. Garden variety pump sprayers are unsuited because their fittings fail in contact with most alkoxysilane-based consolidants. Sprayers with brass, stainless steel, polypropylene (Nalgene), or Teflon fittings are necessary.

Pump sprayers deliver large volumes of liquid and lead to significant waste by uncontrolled runoff (fig. 5.6). Because of their inherently low viscosity, alkoxysilanes have a tendency to run down onto areas not designated for treatment. With smaller objects or smaller treatment areas, delivery devices are also scaled down, thereby improving control. For example, small hand-misters provide good control over runoff, though they also introduce air bubbles that impede penetration into the stone, promote evaporation by nebulizing the liquid, and induce hand fatigue through the continued pumping action. In contrast, squeeze bottles eliminate air bubbles, reduce evaporation, and deliver moderate volumes with reasonable control of runoff. The control is not quite as good as with misters or nebulizers, and squeeze bottles lead to some hand fatigue. Glass or Nalgene pipettes or eye droppers can be used for even smaller-scale delivery. Finally, natural-bristle brushes have excellent control, introduce few air bubbles, limit evaporation and runoff, induce little hand fatigue, and transfer the least amount of consolidant to the stone in a given time. For this last reason, brushes have been used primarily on smaller objects and with uncatalyzed formulations for which the treatment schedule is more relaxed. Along with these more controlled delivery methods, tenting has also been used both for safety and to control evapo-

Figure 5.5.

Several delivery devices are pictured here. The containers and fittings must be stable to the solvent power of alkoxysilanes. The device chosen should match the scale of the project and the need to control runoff.

Figure 5.6.

Runoff with alkoxysilane consolidants can be significant, as shown here (H. Weber 1985).

ration (Bradley 1985), and in outdoor environments treated areas are often covered with aluminum foil (Lukaszewicz 1995). Other methods of application used to improve penetration and/or limit evaporation are vacuum impregnation (Hempel 1976), cotton padding saturated with consolidant (Rossi-Manaresi 1995 and Alessandrini 1988), and, for small objects, partial or total immersion.

A final note on application is the subject of dilution of alkoxysilane formulations. Dilution was first suggested by Laurie (1925) as a means to lower the viscosity of ethyl silicate to ease its penetration into the stone. In more recent literature dilution is suggested to reduce "overconsolidation" of deteriorated surfaces as compared to undeteriorated stone (Ettl and Schuh 1989; Koblischek 1995; Wendler, Klemm, and Snethlage 1991). This dilution is usually achieved with ethanol, although mineral spirits, white spirits, toluene, xylenes, acetone, and methylethylketone also have been used (Emblem 1947).

Consumption

The best means to estimate the amount of consolidant to be used is to execute a mock-up (that is, a test treatment on a small area of the object, monument, or building). Without a mock-up, consumption may be difficult to predict and depends on the method of delivery and the associated runoff. An important factor governing runoff is the geometry and orientation of the surfaces being treated: vertical surfaces engender the most runoff, overhanging horizontal surfaces are next, and up-facing horizontal surfaces least. As shown in figure 5.6, runoff can be significant and can deliver the consolidant to areas where it is neither wanted nor needed. Methods of controlling runoff include the following:

1. Choosing an application technique that discourages runoff.
2. Applying solvents such as acetone or methylethyl-ketone where runoff is predicted. This application limits the access of the consolidant by filling pores and/or dilutes it as it reaches the undesired areas.
3. Dabbing with dry or solvent-bearing paper towels or cloth rags. (This can be a step by itself or as a follow-up to Step 2.)
4. Inserting mechanical devices to direct runoff away from undesired areas.

This last approach is used when there are regular divisions such as mortar joints between sections of stone. Figure 5.7 shows how a joint is cut out and a wooden slat wrapped in plastic sheeting is inserted. The device directs most of the runoff onto the sheeting and away from the stone. Some practitioners have gone so far as to collect and reuse the excess carried away by the sheeting (Wihr 1976).

Other important factors that affect consumption are the permeability and porosity of the stone. A practical means of measuring these properties is to perform a capillary uptake test using protocols such as ASTM C97. From this test the rate of uptake of water (or hydrocarbon solvent or consolidant) and the porosity of the stone relevant for consolidation can be calculated.[27] An on-site variation to obtain the same information is the so-called Karsten tube, illustrated in figure 5.8 (Leisen, Poncar, and Warrack 2000).

For a given area of stone, a reasonable estimate of the amount of consolidant to be used can be calculated by the following:

$$\text{(surface area) x (porosity) x (depth of penetration)} = \text{volume of consolidant}[28]$$

Figure 5.7

This diagram shows how runoff can be controlled more easily on sculptures, monuments, and buildings that have regular divisions between stones such as mortar joints.

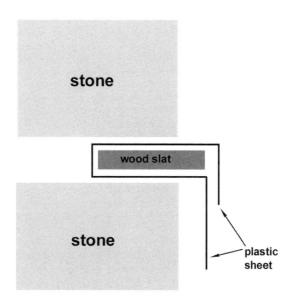

For example: 1 m^2 of surface for a stone of 20% porosity that is saturated in consolidant to a depth of 2.5 cm (1 in) consumes 5 L, that is, 5 L/m^2, or 1 pt/ft^2, excluding runoff. The equation does not account for permeability. Stone with lower permeability imbibes liquid more slowly and, consequently, will have more runoff if the rate of delivery of the consolidant is not reduced.

Not surprisingly, the literature reports a wide range for consumption of alkoxysilanes in the consolidation of stone—from 1 L/m^2 to 40 L/m^2—and there appears to be no consistent relationship between stone type and consumption.[29] Outside of runoff, porosity is the most important factor governing consumption. Stones such as granites and marbles are inherently less porous and require less consolidant. A good working *average* is about 5 L/m^2 or 1 pt/ft^2 (Arnold 1978; Price 1981).

Application Schedule

For consumption of Tegovakon V and Wacker OH on ferruginous sandstone, De Witte and Bos (1992) report that two to four applications were executed. The number of applications influences consumption, but the schedule of these applications is equally important. Treatment schedules have two time frames: within the same day (2–8 hours) or over several days or weeks. In the first time frame, as many as fifteen applications may take place. As suggested above, their number and timing depend primarily on the stone's permeability. The product literature from ProSoCo (*Conservare* consolidants) indicates that applications are grouped in threes, and each group of three applications is referred to as a cycle.[30] They recommend that 5 to 15 minutes lapse between applications within a cycle and that 20 to 60 minutes lapse between cycles. As time passes and applications accrue, uptake slows until the stone can no longer take in consolidant and the treatment is halted. Between applications and, most important, at the end of the treatment, it is necessary to dab areas

Figure 5.8

Karsten tubes are well suited for determining the rate of uptake of water in the field. These data can help to estimate the consumption of consolidant during treatment (H. Leisen 2002).

Figure 5.9

Excess consolidant must be removed with paper towels, cloth rags, and sometimes with solvents. The glazed surface that results from this condition will be difficult or impossible to remove after the consolidant gels and hardens.

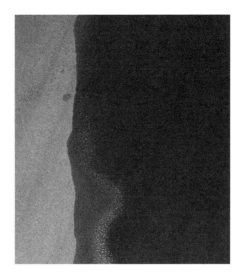 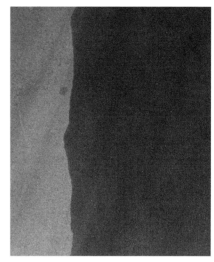

with paper towels or cloth rags to remove excess consolidant. Applying volatile solvents such as acetone assists this removal. It is important to remove all excess, or the surface will acquire a glaze, which is difficult to remove after the consolidant has gelled (fig. 5.9).

This application program is not difficult to execute for small objects, but it can be cumbersome when treating large surfaces. In fact, attempts to treat two such adjacent areas in sequence are quickly confounded. Treating too large a single area at once can also lead to complications. By the time a single application is completed the waiting time for the next application has passed. Furthermore, unless treatment is performed from the bottom to the top, during the second application runoff will pass onto areas not yet ready for treatment. For these reasons, a single operator should not attempt to treat more than 10 m² at a time.[31] A nine-application program—3 cycles of 3 applications—will take approximately three hours, including setup and cleanup. Therefore, a single operator should plan to treat two such areas in a day.

Although in practice this kind of treatment schedule is not always strictly adhered to, there is logic in the program. The rationale is to provide a continuous flow of liquid to the surface (often referred to as wet-on-wet applications) and at the same time give the stone enough time to take in the liquid. Liquid delivered too quickly leads to excessive runoff and waste of consolidant, and too slow a delivery leaves gaps of air in the stone between applications. The goal is to achieve wet-on-wet applications with little or no runoff. If the stone takes in the consolidant quickly, then subsequent applications must take place more quickly, and this reduced time between applications will also require a reduction in the size of the area being treated. It also allows the operator to move onto new sections for treatment in a shorter time span. Conversely, slower uptake allows larger areas to be treated.

Another approach is to carry out applications on successive days. This has the advantage of a more relaxed schedule of treatment and significantly reduces the amount of waste from runoff. With catalyzed formulations, gelation and some shrinkage of the gel occur overnight, leaving space for the next day's application. For uncatalyzed formulations, either neat or with added water and/or solvents, gelation does not necessarily occur overnight. However, in the case of MTMOS much of it can evaporate in a 24-hour period, particularly if limestone is being treated.

This approach has the obvious disadvantage of prolonging the overall time of treatment and, in outdoor settings, of prolonging the period when it is necessary to be concerned about weather. For these reasons it is more often employed indoors. In fact, manufacturers and suppliers of catalyzed formulations do not recommend treating on subsequent days because of concern for premature gelation of later applications.[32] This problem has not generally been noted with OH- and H-style formulations. On the other hand, Brethane has been reported to gel prematurely under these circumstances (Larson 1983).

After application is complete—including solvent wash, dabbing with paper towels or cloth rags (or solvent swabs with small objects [Thickett, Lee, and Bradley 2000]) to prevent "glazing," protection from sun and rain with plastic sheeting or aluminum foil, delivery devices cleaned with solvent to prevent clogging from gelled consolidant—the waiting period begins.

Conflicting Activities

The waiting period is not just for the full curing of the consolidant per se. The conservation of buildings, monuments, and sculpture also involves cleaning, pointing, grouting, plastic repair, stone removal and replacement, salt removal, and application of water repellents. Many of these activities involve water that can interfere with, or be compromised by, the curing of alkoxysilanes. It is necessary to schedule the sequence and the timing of activities to avoid or minimize conflicts.

The most common of the activities named above is cleaning. Cleaning is achieved with materials and techniques that involve water— water misting, waterborne chemicals, and wet abrasives—and those that do not involve water—lasers, solvent-borne chemicals, and dry abrasives. In preparing a conservation program that involves both cleaning and consolidation, two questions must be answered. First, do the conditions of the stone allow for cleaning before consolidation? Second, can the stone be adequately cleaned after consolidation?

Regarding the first question, if granular disintegration, scaling, flaking, or exfoliation is advanced and pervasive, then cleaning with techniques other than lasers will usually result in significant loss of stone. For these conditions the same approach described earlier under *Preconsolidation* is used; that is, an initial, less than full-scale consolidation is carried out. Cleaning then commences, and full consolidation takes place when the stone is sufficiently dry. Some chemical cleaners contain bases such as sodium and potassium hydroxide or hydrochloric, hydrofluoric, and organic acids. These accelerate gelation and therefore must be thoroughly rinsed from the stone before consolidation.

There is no straightforward answer to the second question, and mock-ups must be performed. Often it is difficult to clean light-colored

Figure 5.10

The left part of this marble sculpture was consolidated with *Conservare* OH before cleaning; the right side was not consolidated. Notice that the level of cleaning is greater on the right. The consolidation can limit the level of cleaning as well as the loss of stone. Photo: Chris Gembinski

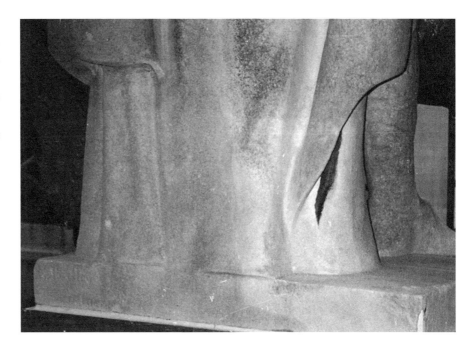

Figure 5.11.

Even with so-called nonhydrophobic consolidants the surfaces of treated stone remain water repellent for weeks or months after treatment. Here, a sandstone treated with *Conservare* OH is still water repellent after twelve weeks. This condition makes it difficult to carry out aqueous cleaning or to install grouts, pointing, or composite repairs that are water-based.

stone such as many limestone and marble with aqueous-based methods after consolidation with alkoxysilanes (fig. 5.10).[33] Nominally hydrophobic consolidants such as Wacker and *Conservare* H, Tegovakon T, RC80 and RC90, and Brethane make aqueous cleaning even more difficult. However, even so-called nonhydrophobic consolidants such as the OHs retain water repellency for long periods after initial treatment. Lukaszewicz (1995) has shown that this period can extend to from several weeks to several months and is dependent on relative humidity (fig. 5.11).[34]

For conflicting activities such as grouting, pointing, and plastic repair, three factors are important in association with alkoxysilane consolidants: (1) the adhesion of grouting, pointing, and plastic repair materials to the stone; (2) the setting and curing of grouting, pointing, and plastic repair materials; and (3) the curing of the alkoxysilane consolidant. Good adhesion depends on both chemical and mechanical bonding of the applied material to the stone substrate. Stone surfaces made hydrophobic by consolidants, if only temporarily,[35] cannot be wetted by materials bearing water. Consequently, both chemical and mechanical bonding are limited.

For such bonding to be achieved, the consolidant must fully hydrolyze, condense, gel, shrink, and lose all of its hydrophobic properties *before* the grouting, pointing, or plastic repair is installed. This may take as long as eight weeks. Felix (1995) has stated that for stones with higher clay contents that may experience shrinkage on consolidation with alkoxysilanes, pointing should take place after consolidation.

The foundation components of grouts, pointing, and plastic repair materials are lime, hydraulic lime, pozzolanic (natural) cement, and Portland cement (Boyle 1980; Hewlett 2001). Water is added at the time of installation, water that initially makes them plastic and that is essential for their chemical and physical transformation from a plastic to solid. Likewise, alkoxysilane consolidants require water for their transformation from liquid to gel. If either is applied too soon on the other, then the competition for water will compromise the curing of both.

Darkening and Color Changes

All stones darken during treatment with alkoxysilanes in the same way that wet stone is darker than dry stone.[36] In some cases the change is permanent, but it is difficult to predict the degree or persistence of the effect from the literature alone. Costa and Rodrigues (1996) showed that dark minerals in granite became darker and feldspars became more yellow with Wacker OH. Bradley (1985) ascribed darkening only to catalyzed treatments, and Price (1981) noted color enrichment with Brethane that lessened after two to three months but did not go away. However, clays and iron oxides treated with various alkoxysilanes darken more than quartz, feldspars, and calcite, even with uncatalyzed MTMOS.

There have been several reports indicating that volcanic rocks darken with Wacker OH, Tegovakon V, and SS-101[37] but that the stone often reverted to its original color in time or after artificial weathering (Nishiura 1987b; Wheeler and Newman 1994). RC90 was found to induce darkening that persisted on both carbonate and silicate substrates, and Verges-Belmin et al. (1991) saw yellowing on marble with black crusts that they ascribed to mobilized organic compounds in the crust. Finally, Rossi-Manaresi (1995) reported whitening that remained for more than twenty years. Such whitening usually occurs when wet stone is treated.

A reliable way to predict the degree and persistence of color changes or darkening is, once again, to perform a mock-up. Small samples can be removed and treated, or a small area of the object, sculpture, monument, or building can likewise be treated. It is frequently difficult to build the necessary time into the schedule of a conservation project to make this kind of assessment.[38] In the field, two approaches have been used to reduce color changes: applying solvents such as acetone or ethanol, followed by dabbing with paper towels or cloths (the same process to eliminate glazing); and performing abrasive "cleaning," which essentially lightens the stone by increasing diffuse reflectance. The latter technique is performed only under the most extreme circumstances.

Figure 5.12

It is difficult to consolidate marble through black crusts. Consolidation is nonetheless desirable because the stone below the crust is frequently friable. Light cleaning with an abrasive or a laser can significantly improve the penetration of the consolidant.

Application on Black Crusts

As Hempel and Moncrieff (1976) pointed out, it can be difficult to consolidate marble through black gypsum crusts. Nonetheless, the stone below these crusts is often friable and in need of consolidation (fig. 5.12). Uptake of the consolidant is generally slow because of the low permeability of the crust, and consequently treatment times are longer. Areas of black crust are usually localized on marble surfaces (as are areas in need of consolidation) so that extended treatment times are more easily justified if the area requiring treatment is relatively small and easily segregated from other activities. Light cleaning before consolidation—preferably by microabrasion or lasers—can increase the rate of the consolidant's uptake. Marble with black crusts that have been consolidated may not clean as completely as untreated marble, and if RC90 is used the toluene in this formulation may mobilize organics in the crust that are redeposited in the marble (Saiz-Jimenez 1991; Verges-Belmin et al. 1991). This effect has not been noted with either the older formulations of OH, which contained acetone and methylethylketone solvents, or the new formulations (OH100), which contain no added solvents.

Water Repellents

The application of water repellents is only performed after consolidation. The active, hydrophobic moieties in most water repellents are alkyl groups directly attached to silicon, including methyl, ethyl, *i*-butyl, and *n*-octyl. The backbone for the alkyl groups consists of siloxanes of various molecular weights and degrees of cross-linking and may themselves contain some reactive alkoxy groups. The medium or diluent is either water or organic solvents such as alcohol or mineral spirits. Waterborne formulations will interfere with the curing of grouts, pointing, plastic repairs, and alkoxysilane-based consolidants. Again, as many as eight weeks must pass before the application of such a water repellent. The same is true for organic solvent–based formulations. Ethanol and reactive alkoxy groups in some water repellents will strip water from the lime and cement and cause incomplete curing. Mineral spirits will trap moisture and also cause incomplete curing.

Organic solvent and alkoxy-reactive water repellents can be applied much sooner over alkoxysilane consolidants (as opposed to grouting, pointing, and plastic repair materials containing lime and cement). In this case, the waiting period can be reduced to two weeks because the water repellent is chemically similar to the consolidant.

A question often posed by conservators, architects, and contractors is, if both consolidation and water repellency are deemed necessary, should this be achieved with hydrophobic consolidants or the nonhydrophobic equivalent followed by a water repellent? The clear advantage of using hydrophobic consolidants such as Wacker H is that only one treatment needs to be performed, resulting in a significant cost- and time-savings on large projects. A second advantage is that water repellency provided by gels rather than coatings has been reported to last longer (Ginell, Kumar, and Doehne 1995). There are disadvantages to this approach. First, as mentioned above, using hydrophobic consolidants prohibits the subsequent application of any water-based treatment: cleaning materials, grouts, pointing, or plastic repairs. Second, water repellents can be applied selectively and locally on top of OH or other nonhydrophobic consolidants (Useche 1994). This local application limits interference of the water repellents with other water-related treatments to the areas where it was applied.

Safety and Storage

Product literature and Material Safety Data Sheets should always be consulted before use, and the precautions laid out therein followed closely. For all formulations, at a minimum, gloves are to be worn for skin protection, particularly for the lead naphthenate catalyst in Brethane and, to a lesser degree, the dibutyltindilaurate catalysts in many other formulations. Where application is less controlled, chemical-resistant overalls are advised. Eye protection is also essential because some compounds such as MTMOS and TMOS can cause clouding of the cornea. Respirators are also necessary; positive air feed varieties are more comfortable and are mandatory if the operator has substantial facial hair such as a mustache or beard. Work areas should be well ventilated during and after application. Another safety note is that plastic sheeting and metal scaffolding become slippery when coated with alkoxysilanes. There should be no open flames when applying alkoxysilanes. Alkoxysilanes should not be discarded in sinks or sewers, and clothes, paper towels, and cloth rags should be allowed to dry in air before discarding. Alkoxysilanes (especially formulations with catalysts) must be stored in dry places. For small quantities, up to 5 gallons or 20 liters, containers can be placed in well-sealed plastic bags with silica gel that has been completely dried.

The information provided in this chapter should be considered guidelines and suggestions. Most of the suggestions have been gleaned from the experiences of practicing conservators. Even a step-by-step approach should be viewed as a recipe rather than as a fixed rule: the basic idea is followed, but the artistry and creativity of the individual always enters into the final result. It is through this interaction of received wisdom and individual creativity that we expand our knowledge and improve our practice, and conservators are essential to this process.

Notes

1. From the product literature for OH100 from ProSoCo, Lawrence, Kansas (www.prosoco.com).

2. For an excellent description of the forms of weathering on stone, see Fitzner, Heinrichs, and Kownatzki 1995.

3. The dislodgement and loss are due to the low viscosity and low surface tension of alkoxysilanes and the fact that the gaps between the stone fragments and the stone substrate are too large to be spanned by either the consolidating liquid or the resulting gel.

4. ACRYLOID B72 is a block copolymer of ethylmethacrylate and methylacrylate; ACRYLOID B48N is a polymer of methylmethacrylate.

5. See also Matero and Oliver 1997.

6. See also Miller 1992.

7. Solvents such as acetone, methylethylketone, and ethanol, when added to some formulations, may alter the solubility of various resins.

8. This work was carried out by Li-Hsin Chang during her internship at the Sherman Fairchild Center for Object Conservation at the Metropolitan Museum of Art.

9. Two recent examples were performed under the guidance of Building Conservation Associates, New York, at the Lord Memorial Fountain in Somerset, New Jersey, and the Toledo Museum of Art in Toledo, Ohio.

10. Funcosil is a line of consolidants by Remmers Baustofftechnik GmbH, Bernhard-Remmers-Str. 13, 49624 Löningen, Germany.

11. HCT is produced by ProSoCo, Lawrence, Kansas (www.prosoco.com).

12. Roedder (1979) also suggested using solutions of ethyl silicate and ethanol to treat wet stone. Hempel and others sometimes referred to this initial application of solvent as "ventilating the pores." Its stated purpose was to ease the ingress of the alkoxysilane either by evaporating the resident water or by forming a solution with it that could allow the homogeneous incorporation of the invading consolidant (Hempel and Moncrieff 1976).

13. Higher temperatures will increase both evaporation and reaction rate. If humidity is low (less than 20%), then the reaction rate will also be slow and the increased evaporation becomes more important than the increased reaction rate at higher temperatures.

14. Hiroko Kariya was an objects conservator in the Conservation Department of the Brooklyn Museum of Art, Brooklyn, New York. This work was performed under the aegis, and with the support, of the Oriental Institute of the University of Chicago.

15. Hiroko Kariya, pers. comm. October 2002.

16. Koestler and Santoro (1988) demonstrated that some consolidants can be a food source for various biological growths.

17. Epilithic lichens grow on the surface of stone; endolithic lichens penetrate below the surface.

18. These approaches are based on the author's field experience.

19. One concern is the formation on salts that may damage the stone.

20. D2 is supplied by Cathedral Stone Products, Inc., Jessup, Maryland.

21. See Butlin, Yates, and Martin 1995; Plehwe-Leisen et al. 1996.

22. The reporting of this effect is actually contradictory. Leznicki (1991) indicates that Wacker H did not inhibit biological growth, while Riederer (1977) indicates that it does.

23. For an illustration of this problem, see Mangio and Lind 1997.

24. Hempel and Moncrieff 1972, 1976; Larson 1983.

25. Long-term poulticing (over a few days) can lead to biological growth under the paper pulp and on the stone surface.

26. Saleh et al. 1992a, 1992b; Rager, Payre, and Lefevre 1996.

27. Values for porosity depend on the technique used to make the measurements. The most common techniques are capillary water uptake, mercury intrusion porosimetry, and BET nitrogen. The latter two techniques generally find more porosity because they detect smaller pores that cannot be reached by capillary water or, for that matter, by the consolidant. Therefore, capillary water uptake is a more relevant measurement of porosity for estimating consolidants consumption.

28. See Price 1981 for a review of this calculation.

29. For example, Galan and Carretero (1994) report 2.4 L/m^2; Lukaszewicz (1995), 5 to 15 L/m^2 for sandstones with 5 to 20%; Weber (1976), up to 20 L/m^2; and Wihr (1978), up to 40 L/m^2.

30. The conceptual framework and terminology for this practice were originally developed by Weber (1985).

31. Price (1981), in fact, suggests application areas of not more than 1 m^2.

32. Weber (1977) suggests applications of Wacker OH over a 1- to 2-week period.

33. Larson (1983) stated that such stones could only be cleaned with abrasives after consolidation with MTMOS, and Hanna (1984) indicated that it was difficult to clean limestone after consolidation with OH.

34. The water repellency of TEOS-based consolidants derives from residual ethoxy groups. These ethoxy groups can be lost over time by continued hydrolysis. Some additional repellency applies for those consolidants that contain solvents such as MEK or mineral spirits. The water repellency from these more volatile components is generally more short-lived than the repellency from residual ethoxy groups.

35. Clearly, if a hydrophobic consolidant such as H is applied, then these repair materials cannot be installed at all.

36. When air that normally fills the pores is replaced with a liquid of higher refractive index, the amount of reflection is reduced and the stone appears darker.

37. This consolidant is produced by COLCOAT, Ltd., Tokyo.

38. The author treated a scoria chair by Scott Burton for which the stone did not return to its original color for one year. On the other hand, the green Nova Scotia sandstone for Bethesda Terrace in Central Park in New York has not reverted to its original color after twenty years in a location where runoff was concentrated.

Chapter 6
Laboratory and Field Evaluation of Service Life

This chapter addresses the question of how long an alkoxysilane consolidant treatment is expected to last. The literature provides information in two forms: laboratory testing and field evaluations. Laboratory experiments are usually designed to *differentiate* among competing or simultaneous processes. This differentiation is achieved by fixing some influences, varying others, and measuring the outcomes. In most field studies all influences are controlled by nature, and the resulting evaluation *integrates* their input. Differentiation allows for easier interpretation of the data in a limited context. Integration takes all factors into account, but the data are more difficult to interpret because the relationship between cause and effect can often be established only tenuously. The point of departure for this chapter is the field study published by English Heritage titled "Stone Consolidants: Brethane Report on an 18-Year Review of Brethane-treated Sites."[1] The study is a model of both richness of information and persistence of observation.

This recent work by Martin et al. (2002), which was foreshadowed by Butlin and colleagues in 1995 and Martin in 1996, reports on visits to ten sites over approximately a twenty-year period where treatments were carried out primarily with Brethane.[2] The most recent visits by a member of the survey team were in 1997, and this author visited nine of the ten sites in October 1996 with the cooperation and assistance of English Heritage staff. Table 6.1 lists the sites and provides basic information about the stones and information gleaned from the survey that are pertinent to the question of service life.

As can be seen in the table, the sites contain a good representation of sedimentary rocks often employed for sculpture, monuments, and buildings—ranging from the essentially calcite-pure clunch[3] from St. George's Temple at Audley End and the biosparite[4] at Berry Pomeroy to the heterogeneous, clay-bearing, calcareous sandstone at Goodrich Castle and the quartz wacke or arenite at Rievaulx Abbey. The survey team used seven criteria to evaluate the performance of Brethane: color, degree of soiling, degree of biological growth, degree of decay relative to untreated stone, degree of powdering, degree of scaling, and degree of water repellency. As the authors note, the survey is hampered by the lack of good notes on the conditions of the stone at the time of the original consolidation treatment.

Table 6.1

Data on the stones and conditions for the Brethane Survey

Site	Stone Type	~ Carbonate Content	~ Silicate and Iron Oxide Content	Porosity	Years to General Level of Decay: Treated = Untreated	Years to Level of Powdering: Treated = Untreated	Years to Level of Scaling: Treated = Untreated	Years to Level of Water Repellency: Treated = Untreated	General Comment
St. George's Temple: Audley End	Clunch	100%		< 25%	> 20	> 20	< 4	> 20	after 21 years treated stone is better than untreated stone
Berry Pomeroy	Biosparitic limestone or biosparite	100%		moderate–high (20%?)	9–15	> 14	< 5	> 14	after 5–10 years treated and untreated stone are similar
Howden Minster	Dolomitic limestone or dolomite	100%		25%	< 3	> 13	8-13	> 13	after 13 years treated and untreated stone are similar
Chichester Cathedral	Glauconitic, siliceous limestone	< 100%	> 0%	3–5% (?)*	> 19	> 19	> 19	8–10	after 19 years treated stone is better than untreated stone
Bolsover Castle (1)	Dolomitic, sandy limestone	82%	18%	29%	~ 15	~ 15	< 6	6–10	after 15 years treated stone is better than untreated stone
Goodrich Castle	Calcareous sandstone	12%	88%	24%	~ 10	> 19	8-10	>19	after 19 years treated stone is slightly better than untreated stone
Bolsover Castle (2)	Lithic sandstone	n.d.	n.d.	4.5% (?)*	> 15	> 15	10–15	> 15	after 15 years treated stone is better than untreated stone
Kenilworth Castle	red sandstone	2%	98%	26%	4–8	0–8 (?)**	13	> 13	after 13 years treated stone is slightly better than untreated stone
Rievaulx Abbey	Quartz arenite or quartz wacke***		100%	> 40%	~ 13	> 18	5-6	8–13	after 18 years treated stone is slightly better than untreated stone
Tintern Abbey	Lithic graywacke		100%	29%	8-10	> 20	10-20	9–11	after 20 years treated and untreated stone are similar
Sandbach Crosses	red silica sandstone		100%	16–20%	n.d.	n.d.	n.d.	> 21	treated stone is in excellent condition

Notes:
*Presumably, these are the porosities of the undeteriorated stones.
**The survey gives conflicting information for this category at Kenilworth Castle.
***Analysis indicates this may be an arenite or a wacke, depending on the amount of matrix present in any given sample. A sample analyzed by Robin Sanderson for English Heritage was identified as quartz arenite.

The first three categories of evaluation—color, degree of soiling, and degree of biological growth—are of limited value in the context of this chapter and problematic even in the context of the survey. For the survey the only acceptable condition for color is that treated and untreated stone are identical. Clearly, a change in color due to the treatment is undesirable (see chaps. 3 and 5). However, by the survey's criteria the following scenario would also be considered undesirable: a stone remains the same color as untreated stone at the time of treatment and over time becomes darker than untreated stone. On the face of it this might be an appropriate judgment. In some cases, however, treated stone may appear darker because it is more resistant to erosion.[5] The continued erosion of untreated stone ensures that it remains lighter. Similar logic applies to soiling: it may build up because treated surfaces resist erosion due to the positive effects of consolidation. Finally, biological growth is closely tied to general environmental conditions—temperature, relative humidity and rainfall, and available nutrients—and patterns of runoff on structures that produce persistent wetness. All of these conditions are generally unrelated to the presence (or absence) of a consolidant. For that matter, a consolidant may be performing its function adequately despite the presence of biological growth.[6] Therefore, emphasis is placed here on the other categories of evaluation: degree of decay relative to untreated stone, degree of scaling, and degree of powdering, with somewhat less weight placed on degree of water repellency. With these caveats in place, let us return to table 6.1 to learn more about the service life of Brethane.

The table is organized with limestones at the top in descending order of carbonate content and continues with sandstones in ascending order of silicate content. For limestones only the coarse-grained dolomite at Howden Minster did not fare well with treatment: according to the surveyors, in less than three years the general level of decay of treated and untreated areas was similar. This stone was relatively friable and deteriorated when I visited in 1996, and there appeared to be little difference in treated and untreated areas. The clunch at St. George's Temple and the siliceous limestone at Chichester Cathedral are still doing well after twenty years. In the case of Chichester, untreated stone was also in relatively good condition in 1996. The treated biosparitic limestone at Berry Pomeroy has lasted ten to fifteen years before becoming similar to the moderately deteriorated untreated stone; and the treated dolomitic limestone[7] at Bolsover Castle lasted fifteen years despite the continued presence of salts.[8] Thus when a moderate level of deterioration is noted—St. George's Temple, Chichester Cathedral, Berry Pomeroy, and Bolsover Castle—the treatment appears to be effective for at least ten years and probably longer. These four limestones also have relatively small grain sizes and small pore spaces. For the higher level of deterioration at Howden Minster, the treatment was not very effective, and additional treatment was probably warranted after three years. This stone also has larger grain sizes and larger pore spaces than the other four limestones in the survey.

Since alkoxysilane consolidants are unable to secure most flakes and scales because the gels cannot bridge gaps larger than 50 μm (Wendler

Figure 6.1

Heavy deposits of magnesium sulfate
salts can be found on the surface of the
dolomite at Bolsover Castle. These salts are
the product of the interaction of sulfur
dioxide in the atmosphere and the minerals making up the stone. Consolidation
cannot prevent damage to the stone
caused by the continuing presence of
these salts.

et al. 1999), evaluating Brethane's performance on scales is also problematic; it is unlikely to score well for a function it is unable to perform. At
St. George's Temple scales of less than 10 cm were recorded in the survey
four years after the initial treatment with Brethane, and this condition
persisted but did not worsen over the next twenty years. It is tempting to
guess that these scales were present at the time of the initial treatment. At
Berry Pomeroy (biosparitic limestone or biosparite) no scales were noted
until five years into the survey, and this condition worsened over the next
fourteen years. For the siliceous limestone at Chichester Cathedral, no
scales were present, and none formed in nineteen years. At Bolsover Castle
(dolomitic limestone) significant scales appeared in less than six years.
The formation of scales at this site may not represent the failure of the
treatment per se but rather the failure to remove a major source of
deterioration—magnesium sulfate salts (fig. 6.1).[9]

It is worth noting the degree of water repellency exhibited over
time for the limestones listed in table 6.1. For the siliceous limestone at
Chichester Cathedral and the dolomitic limestone at Bolsover Castle,
water repellency vanished in less than ten years. For Howden Minster,
Berry Pomeroy, and St. George's Temple, it is still functioning, although
at significantly reduced levels, after at least thirteen, fourteen, and twenty
years respectively.

Turning now to those sites constructed with sandstones, these
stones range from the calcareous sandstone at Goodrich Castle that contains upwards of 10% calcite to stones that contain essentially quartz,
feldspars, iron oxides, and clay minerals. Looking at the category of general
level of decay of treated versus untreated stone, the treated lithic sandstone
at Bolsover Castle appears to be performing well after fifteen years. The
arenite or wacke at Rievaulx Abbey is so highly deteriorated that its porosity exceeds 40%. Salts are active at the location of treatment, as indicated
by increasing levels of efflorescences (fig. 6.2), and consolidated areas are
performing only marginally better than untreated areas after eighteen years.
This stone probably warranted retreatment in less than eight years. At

Figure 6.2 (left)

Increasing levels of salt efflorescences are present at Rievaulx Abbey. Whereas consolidation can strengthen stone, it cannot prevent damage by salt crystallization.

Figure 6.3 (right)

Persistent rising damp at Kenilworth Castle has carried salts into the lower courses and has caused significant damage to the sandstone.

Figure 6.4 (left)

Scales and flakes are prevalent at Rievaulx Abbey. These forms of deterioration are not easily stabilized by alkoxysilane consolidants.

Figure 6.5 (right)

Scales and flakes are abundant on the stone surfaces at Goodrich Castle. Stabilization lasted five to ten years.

Goodrich Castle it took approximately ten years for treated and untreated calcareous sandstone to appear similarly decayed, and, a similar time frame applied to the siliceous sandstone at Tintern Abbey. At Kenilworth Castle, where rising damp and salts are major problems, as evidenced by highly deteriorated stonework and extensive efflorescences in the lower courses (fig. 6.3), the treatment functioned only four to eight years.

For the category of degree of scaling in relation to these sandstones, Brethane performed least well at Rievaulx Abbey, where significant scaling has been present from at least 1981 (fig. 6.4). The gaps between these scales and the stone substrate are simply too large to be bridged by the Brethane gel and therefore cannot adequately be stabilized by this treatment. At Goodrich Castle, where scales and flakes are now prevalent (fig. 6.5), no scales were noted until eight years into the survey. For the sandstone at Bolsover Castle, scales were noted in ten to fifteen years. Surprisingly, for Tintern Abbey, with its large and extensive spalls (fig. 6.6), Brethane held up for ten to twenty years, probably due to the larger areas of attachment of the spalls to the stone substrate.

Figure 6.6

The large spalls at Tintern Abbey were stabilized well by treatment with Brethane. Unlike scales and flakes, spalls often have large areas of attachment to the stone substrate. This condition is more conducive to treatment with alkoxysilane consolidants.

Finally, water repellency was generally preserved on these sandstone substrates. Loss occurred at approximately ten years at Rievaulx and Tintern Abbey and at thirteen to close to twenty-one years at other sites.

The survey also evaluates Brethane under the category of degree of powdering, which represents the degree to which powder or actual pieces of stone are removed by lightly drawing a finger across the surface. The survey found that the only two stones that do not perform well in this category are the sandstone at Kenilworth Castle and the dolomite at Bolsover Castle.[10]

Martin et al. (2002) draw several conclusions from this study. Those that are most important for our purposes are: (1) ten to fifteen years after application, stonework should be visually indistinguishable from an untreated area but should be weathering at a reduced rate; and (2) Brethane is more successful when applied to limestones than sandstones. The first conclusion offers a rule of thumb for the service life of Brethane, and perhaps other alkoxysilane consolidants. Martin et al. characterize the second conclusion as surprising and unexpected because it runs counter to the traditional wisdom that alkoxysilane consolidants work better on sandstones, for which they have chemical affinity. This conclusion, and the hypotheses they offer to explain it, warrant closer examination.

The hypotheses offered to explain the better performance on limestones are: (1) it is due to the limestones' general higher porosity and higher saturation coefficients; and (2) it is due to the different decay mechanisms associated with limestones and sandstones, chemical versus mechanical. Table 6.1 demonstrates that the porosities of the limestones range from a low of 3 to 5% at Chichester Cathedral to a high of 29%. Excluding the unusually low value for the Chichester limestone, which is undoubtedly for undeteriorated stone, and ascribing 20% to the limestone from Berry Pomeroy, whose porosity Martin et al. (2002:6) called "moderately high," then the average porosity for the group of limestones is 25%. Following the same scheme for the sandstones, that is, excluding the unusually low value for the Bolsover Castle sandstone (4.5%), then the average porosity of the sandstones is 27%. Therefore, the first hypothesis that the performance on limestone is due to their higher porosity is not supported by the data.[11] It is impossible to evaluate the statement about the relative importance of saturation coefficients for these stones, as no data are provided.

The second hypothesis for Brethane's better performance on limestone rests on the supposition that its ability to make surfaces water repellent reduces the chemical damage by acid rain and/or dry deposition. This hypothesis, however, is also not supported by the data. At Howden Minster fewer than three years passed before the general level of decay on treated and untreated dolomitic limestone was the same. Yet the water repellency is still retained after thirteen years; that is, water repellency is preserved and the stone is not. Conversely, at Chichester Cathedral the siliceous limestone lost its water repellency in eight to ten years, but the treated stone is still in better condition than untreated stone at nineteen years and beyond; that is, the stone is preserved and water repellency is not. Clearly, there is no correlation between water repellency—the deterrent to the chemical deterioration produced by acid rain and dry deposition—and state of preservation for these limestones.

Additional and useful information can be mined from the Brethane survey and table 6.1 by posing the following questions: Where did Brethane perform poorly, and why? Where did it perform well, and why?

For sandstone Brethane performed least well at Rievaulx Abbey and Kenilworth Castle. At each of these sites significant sources of deterioration were not removed. Both sites exhibit salt efflorescences that appeared to increase after treatment, and at Kenilworth Castle rising damp has continued unabated (fig. 6.7).[12] In addition, it is reasonable to assume in both cases that the stone was in poor condition at the time of treatment, given the general conditions of untreated stone.

Brethane performed best on the silica sandstone of the Sandbach Crosses and on the lithic sandstone at Bolsover Castle. At Sandbach the relative condition of treated versus untreated stone cannot be assessed because no part of the stone was left untreated. At Bolsover, the small pore spaces (and therefore proximity of the grains) and the dominant silicate mineralogy make this stone a good candidate for treatment, with Brethane (fig. 6.8).

For limestone Brethane performed least well at Howden Minster. This stone has relatively large grain sizes and large, pocketlike intergranular spaces (fig. 6.9). On other limestones Brethane failed to prevent scales from forming or did not stabilize scales and spalls at St. George's Temple and Berry Pomeroy and the dolomite at Bolsover Castle. Brethane

performed well on the limestone at Chichester Cathedral, a stone that is
in relatively good condition.

    With these observations, the conclusions of Martin et al. can be
refined.

- Stones that were in relatively good condition early on in the
  survey and that were probably in relatively good condition at
  the time of initial treatment have stood the test of time well.
  Examples are the limestone at Chichester Cathedral and the
  sandstone of the Sandbach Crosses.
- Stones in generally poor condition at the beginning of the survey (and probably at the time of treatment) degraded quickly.
  Examples are the dolomitic limestone at Howden Minster and
  the sandstones at Kenilworth Castle and Rievaulx Abbey.
- When significant sources of deterioration are not removed,
  treated stone does not perform well: (1) the persistent rising
  damp at Kenilworth Castle and the salts it carries into the
  stone and eventually to the surface produce forces of deterioration that cannot be resisted by any consolidation treatment;
  (2) salts have caused deterioration that has been only weakly

Figure 6.8

Sandstone from Bolsover Castle in plane-
polarized light at 100x magnification: the
small and nearly linear pore spaces and
grain boundaries and the proximity of the
silicate mineral grains that make up about
88% of this stone make it a good candidate for consolidation with Brethane.

scale = 0.2 mm

Figure 6.9

Dolomite from Howden Minster in plane-polarized light at 100x magnification: grain sizes are large (on the order of 0.15–0.2 mm), as are the pocketlike pore spaces. The sizes and shapes do not make this stone a good candidate for consolidation with alkoxysilanes.

checked by consolidation treatment at Rievaulx Abbey; (3) salts in the dolomite at Bolsover Castle have also caused deterioration that is not prevented by consolidation.

- At sites where scaling is a dominant form of deterioration, Brethane does not stabilize it very well. Examples are Audley End, Berry Pomeroy, Bolsover Castle (dolomitic limestone), and Rievaulx Abbey.

At this point other less systematic and less complete field evaluations can expand our understanding of service life. Two sites are particularly helpful in examining the treatment of scaling and flaking more closely. The first is known as Bethesda Terrace, built in Central Park in New York in 1870.[13] The green-to-tan, clay-bearing sandstone from which the terrace is constructed exhibited scaling, flaking, and granular disintegration (fig. 6.10) in many locations and was treated in 1984 with Brethane.[14] Periodic examination of the site over the past twenty-one years reveals that the Brethane treatment continues to stabilize all three forms of deterioration.

The second site, which has been examined (and treated) periodically over the past twenty years, is the Fuentidueña Apse at the Cloisters, a branch of the Metropolitan Museum of Art in New York (Kimmel and Wheeler n.d.). The apse is part of a twelfth-century church constructed from a porous, coarse-grained dolomite that contains clay and quartz (fig. 6.11).[15] A study of the apse in 1983–86 noted scaling, flaking, spalling, and granular disintegration and identified salts present in the stone.[16] In 1987 about 20% of the stone was treated with *Conservare* OH, and by 1992 additional treatments were required, at times on the same stones. Treatments were also carried out in 1993, 1994, and 1995 (Kimmel and Wheeler n.d.).

What is striking about these two sites is that while they both exhibit scaling and flaking, these conditions were stabilized by alkoxysilane treatment for the sandstone at Bethesda Terrace but not for the dolomite at the Fuentidueña Apse. At the English Heritage sites, Brethane was unable to stabilize scaling and flaking wherever it appeared. Two of those sites, St. George's Temple and Berry Pomeroy, are constructed of

Figure 6.10

The scaling, flaking, and granular disintegration at Bethesda Terrace is still stabilized by the Brethane treatment after
twenty-one years. The portion of the
stone visible in the photograph is approximately 18 cm wide.

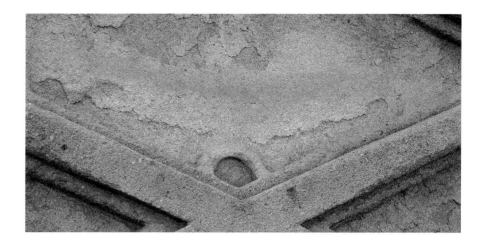

limestone, and alkoxysilanes simply do not have the ability to adhere
scales and flakes to the underlying substrate because of their carbonate
mineralogy. The stone and its conditions at Bolsover Castle are in many
respects similar to the stone and the conditions at the Fuentidueña Apse.
They are both dolomitic limestones with some silicate minerals, and both
sites continue to exhibit salt efflorescences. On the Fuentidueña Apse the
scales and flakes are well removed from the substrate (fig. 6.11; note the
dark shadow cast by the lifting fragments of stone), negating all possibility that they could be secured by alkoxysilane gels.

It is instructive at this point to compare two sites constructed
of sandstone: Rievaulx Abbey from the Brethane survey and Bethesda
Terrace. Although alkoxysilanes may adhere to the dominant quartz and
feldspars for the Rievaulx Abbey stone, figure 6.4 shows that the gaps
between the substrate and the scales and flakes are still too large to be
bridged by alkoxysilane gels. Furthermore, salts continue to be active,
resulting in further deterioration. For the sandstone at Bethesda Terrace
these scales and flakes (see fig. 6.10) lie close enough to underlying stone
that they are well adhered to it; that is, there is enough area of contact,
and the gaps are small enough between them. No salts are present in the
stone at Bethesda Terrace.

Figure 6.11

The porous, coarse-grained, and heterogeneous dolomite used for the Fuentidueña
Apse is susceptible to scaling, flaking,
spalling, and granular disintegration. By
2003 the area of scaling and flaking in the
center of the image was fully detached. The
entire portion of the stone visible in the
photograph is approximately 60 cm wide.

Another important field evaluation of consolidated stone at three national parks was recently published by Oliver (2002). The first site is El Morro, a large natural outcrop of Zuni sandstone, a soft material that was regularly inscribed by visitors for at least the past three centuries. The stone "is composed mainly of uniformly sized quartz sand grains with a lesser component of feldspars. Clay minerals, primarily kaolinite and chlorite, are the only form of cement throughout much of the stone" (Oliver 2002:40). The stone exhibited large-scale exfoliation and both flaking and granular disintegration. Sources of deterioration include rain and rapid drying, extremes in temperature between shaded elevations and those in bright sun, and freezing water. Given these natural environmental conditions, it is surprising that many inscriptions appeared intact in 1993 and at that time showed little change from photographs taken in 1950. As Oliver points out, however, some of these inscriptions in good condition in 1993 were nearly obliterated by 1996,[17] and localized ethyl silicate consolidation[18] was performed in 1996 and again in 1997 to address the losses occurring by granular disintegration. By 2000 more treatment was required to stabilize further deterioration. This is a short time between treatments, and two factors may contribute to the poor performance. The first factor is environmental conditions. Rainfall coupled with robust winds produces strong stresses with wetting and drying (Snethlage and Wendler 1991), and frequent excursions below 0°C produce freezing damage. The second factor is the nature of the stone itself. Given its softness, evidenced by the ability of visitors to casually leave their marks, there must be a relatively large amount of clay matrix, and therefore clastic grains of quartz and feldspars have little or no contact. As weathering of the stone advances and more matrix is lost, the remaining grains of quartz and feldspars are too far apart to be bridged by alkoxysilane gels. The result is weak stabilization of the prevalent granular disintegration.

The second site studied by Oliver is the Square Tower boulder at Hovenweep National Monument in Utah. This boulder is a "very porous, calcite-cemented sandstone, composed of well-sorted, sand-sized grains" (Oliver 2002:41). According to Oliver, the stone is better cemented,[19] deterioration less severe, and the climate more mild than at El Morro. The stone exhibits erosion and granular disintegration (and "friability")[20] caused by uptake of "brackish, sulfate-laden groundwater" (p. 41) and was treated in 1996 and 1997 with an ethyl silicate consolidant. After five years the deterioration is slowed, despite the continuing rising damp, and consolidation was considered relatively successful.

The last site is a column in the convent of Mission San José y San Miguel de Aguayo in San Antonio, Texas. The column, erected in 1861, was made of an "impure limestone composed of a high percentage of angular grains of quartz in a cryptocrystalline calcite matrix" with "lesser amounts of clays and iron oxides" (p. 42). By the 1990s, areas of the stone were very friable, with extensive microcracking and flaking. Deterioration was attributed to gypsum mortars used to set the column drums. In 1992 ethyl silicate consolidation was preceded by reattachment of large flakes with ACRYLOID B72 acrylic resin and followed by application of methyltrimethoxysilane as a water repellent. Loss compensation

Figure 6.12

Large flakes and spalls on the column at the convent of Mission San José y San Miguel de Aguayo could not be treated with an alkoxysilane consolidant. Instead, they were adhered with acrylic resin, followed by full consolidation with ethyl silicate. Large gaps that remained were injected with hydraulic lime grout and other areas of loss also repaired with specially prepared hydraulic lime mortars. Photo courtesy Architectural Conservation Laboratory, University of Pennsylvania

and cosmetic repairs were performed with hydraulic lime mortars. The water repellent was reapplied in 1996, and no deterioration of the column or the repairs was noted up to 2002.

What do Oliver's field observations add to our understanding of the service life of alkoxysilane consolidants? Consolidation did not stabilize granular disintegration at El Morro for even a few years. At this open site the large natural forces of deterioration could not be removed, so any consolidation will have a limited lifetime. The presence of clay in the matrix of the stone limits the strength increases that might help to resist deterioration (Sattler and Snethlage 1988). The extent of the matrix reduces the contact of quartz and feldspar grains, whereas greater contact provides sites that can be bridged by the consolidant. In sum, it is both the conditions of the site and the stone that subvert the consolidation treatment. At Hovenweep National Monument consolidation stabilized the stone at least moderately well for five years despite the salt-laden water continuing to percolate through the stone. This relative success might be attributed to the close-packed nature of the stone, which is conducive to consolidation with alkoxysilanes. The conservation of the siliceous limestone column at the convent of Mission San José y San Miguel de Aguayo offers instruction on the limitations of alkoxysilane consolidation treatments.[21] Matero and Oliver (1997) recognized that the large flakes could not be secured with ethyl silicate (fig. 6.12).

Therefore, these flakes were first adhered with acrylic resin, and additional large gaps between the flakes and the substrate were filled with hydraulic lime grouts. This same approach is now being used on the flakes and spalls on the Fuentidueña Apse: adhesive injection to secure the flakes and spalls, ethyl silicate consolidation to strengthen the stone against the forces of deterioration, and grout injection to fill gaps to support the flakes and to ensure that rainwater moves across and over these flakes and spalls rather than into the gaps.

The collection of field evaluations presented above has provided a wealth of information on the service life of alkoxysilane consolidants that I now supplement with laboratory evaluations. Returning to Martin et al.'s report on Brethane offers a segue to these evaluations, in particular, their statement that Brethane works better on limestone than on sandstone and their surprise at this conclusion. That their hypotheses to explain this conclusion are not supported by their data does not mean that the conclusion itself is incorrect. Their evaluations indicate that the performance of Brethane on limestone and sandstone is nearly the same. This *is* surprising and calls for a closer look at other work that either implies or directly states better performance on sandstone. Table 6.2 is a good example. It compares the percent increase in modulus of rupture (MOR) imparted by various TEOS-based consolidants to Monks Park limestone and to Ohio Massillon sandstone. Clearly, the increases to the sandstone—averaging about 220%—are far greater than the increases for limestone—averaging approximately 52%. However, these *percentage* increases do not necessarily reflect the ability of consolidated stone to resist deterioration. The *actual* MOR values resulting from treatment, shown in table 6.2, provide a different perspective. The average MOR for treated limestone is 9.4 MPa and for treated sandstone 8.5 MPa. By this measure the treated limestone may better resist deterioration because of its higher modulus of rupture.

It is not consistently true in MOR testing that actual values for sandstone and limestone are so similar. Wheeler, Fleming, and Ebersole (1992) used Indiana limestone and Wallace sandstone and found that

Table 6.2

Modulus of rupture and % increase in modulus of rupture for Monks Park limestone and Ohio Massillon sandstone treated with TEOS-based consolidants

Treatment	Limestone		Sandstone	
	MOR (MPa)	% Increase	MOR (MPa)	% Increase
Untreated	6.2 ± 0.6		2.6 ± 0.1	
OH	10.4 ± 0.4	67.5	9.0 ± 0.5	242
TEOS sol/TEOS	9.6 ± 1.1	55.3	9.4 ± 1.0	257
TEOS sol/ethanol	8.8 ± 0.5	42.2	7.3 ± 0.3	179
TEOS sol/n-butanol	9.3 ± 0.6	50.0	7.9 ± 0.5	200
TEOS sol/1,1,1 trichloroethane	9.2 ± 1.2	49.4	9.3 ± 0.3	255
TEOS sol/ methylethylketone	9.5 ± 1.2	53.7	8.8 ± 0.4	234
TEOS sol/acetone	9.1 ± 1.0	47.9	7.7 ± 0.4	194

both MTMOS-based and TEOS-based consolidants produced much higher MORs for sandstone (22.4 MPa and 26.1 MPa) than for limestone (16.8 MPa and 13.6 MPa). These findings underscore the importance of testing and evaluating individual stones and of avoiding generalizations about the performance of alkoxysilane consolidants on particular stone types.

Other examples of laboratory testing shed light on service life. Sattler and Snethlage (1988), in fact, attempt to connect the field to the laboratory. They extracted large cores (8 cm in diameter by at least 50 mm in depth) from buildings that several years before had been consolidated with "silicic acid ester" (presumably Wacker OH). One example is the Seekapelle in Bad Windsheim, a fourteenth-century church built with Schilfsandstein[22] and treated in 1979. The cores were sliced in 5 to 10 mm thicknesses perpendicular to the main axis of the cylinder and the slices subjected to biaxial flexure testing following the protocol of Wittman and Prim (1983). The goal of the testing was to determine both "the quality and the durability of a consolidation treatment" (Sattler and Snethlage 1988:953). Four cores were removed nine years after the 1979 treatment with the following results: (1) two cores show good strength increases and a smooth profile of strength with depth; in the authors' estimation these are conditions of a successful and durable treatment; (2) one core exhibits good strength increases at the surface, with a sudden decrease below the surface; they consider this the result of inadequate penetration of the consolidant at the time of the initial treatment; (3) one core shows insufficient consolidation or the breakdown of the treatment.

Another example is St. Peter's Church in Fritzlar that was constructed with Bundsandstein and treated in 1979. Cores were similarly extracted and subjected to the same testing protocol. In addition, the sandstone was treated in the laboratory at the time of testing to compare the results to the samples taken in the field. The lab-treated samples exhibit good strength increases and a smooth strength-depth profile. After nine years of exposure, field samples have strength increases similar to samples recently treated in the laboratory, but the field samples also exhibit insufficient depth of penetration: there is a sudden decrease in strength at a depth of 10 mm.

Stadlbauer et al. (1996) make a similar connection between the field and the laboratory—this time on the Baumberg sandstone (actually a sandy limestone)[23] used at Clemenswerth Castle. Consolidation treatments had been carried out in 1975 with Wacker VP 1301, the precursor to Wacker H. In 1996 they performed drill resistance measurements (Wendler and Sattler 1996) in the field to determine the strength-depth profile for treated areas. Some of these areas showed scaling prior to the 1975 treatment. Of the six positions tested by drill resistance, nearly all exhibited even strength-depth profiles and sufficient strength. In the area of scaling there was a large dropoff in strength at a depth of 0 to 3 mm.

Others developed connections between the laboratory and the field by constructing testing protocols to simulate field conditions. A good example is the work of Ginell, Kumar, and Doehne (1995). To establish the testing protocol, they first carried out environmental monitoring at the Mayan site at Xunantunich, Belize. With these data in hand, they performed daily cycles between 21°C /85% RH and

44°C /25% RH on both untreated samples of the highly porous (27–56%), fine-grained (1–10 µm) limestone and samples treated with *Conservare* H. The means of evaluating the performance of the consolidant was a microabrasion technique originally developed for plasters by Phillips (1982). Microabrasion was performed: (1) after the consolidant had cured but before temperature/humidity cycling; (2) after 30 days of cycling; and (3) after 180 days of cycling. Before cycling, the loss to H-treated limestone was only 9% of the amount abraded from untreated stone. After 30 days, the loss relative to untreated stone rose to 15% and after 180 cycles to 47%.

Caselli and Kagi (1995) subjected samples of Bacchus Marsh sandstone, consisting of "small angular grains of quartz in a clay-rich matrix" (p. 121), to both natural (outdoor) and artificial (QUV with condensation) weathering.[24] Samples were treated with Wacker OH, Wacker OH with Wacker 280 water repellent, Wacker H, and Brethane. The evaluation of treated and untreated samples was made using Brinell hardness tests.[25] Samples exposed outdoors showed no decrease in Brinell hardness; in fact, slight increases were noted with OH, H, and Brethane. Artificial weathering produced decreases in hardness for OH, OH + 280, and H, and a slight increase with Brethane. In all cases, the surfaces of treated and weathered samples were more than 100% harder than untreated and weathered samples.

Nishiura, Fukuda, and Miura (1984) also performed artificial weathering in the form of freeze-thaw cycles on samples of porous tuff treated with SS-101, a Japanese product similar to the hydrophobic consolidant Wacker H. Like Ginell and colleagues (1995), they established laboratory testing that related to conditions in the field. Evaluation was carried out by performing ultrasonic velocity measurements at 0, 5, and 10 cycles of freezing and thawing.[26] After 10 cycles the ultrasonic velocity of the untreated sample fell by 67%; the SS-101-treated sample, by only 4%.

Felix and Furlan (1994) subjected molasse that had been consolidated with Wacker OH to cycles of wetting and drying and determined changes to both the elastic and rupture moduli using biaxial flexure testing. In contrast to the rather good results of Caselli and Kagi (1995) and Nishiura and colleagues (1984), in only 4 to 6 cycles the elastic modulus of the treated molasse decreased by 86% and the rupture modulus fell 56%. Wheeler[27] also found significant losses (> 50%) to the elastic modulus of Portland sandstone[28] treated with both *Conservare* H and *Conservare* OH in 10 cycles of either wetting and drying or freezing and thawing. It is noteworthy that the changes in ultrasonic velocities for both wetting and drying and freezing and thawing were independent of the water-repellent properties of the consolidant. What might be argued in the case of Nishiura et al. is that the water-repellent consolidant performed well in freeze-thaw cycling because it excluded water and limited freezing damage. For Wheeler, however, both the hydrophobic and non-hydrophobic consolidants performed equally poorly.

Rodrigues et al. (1996) exposed treated and untreated granite to temperature and humidity cycling as well as long-term immersion in water and determined changes in ultrasonic velocity and bending strength. The ultrasonic velocity rose from 2150 m/sec. to 3750 m/sec. on treatment with an unnamed ethyl silicate product. It subsequently fell to 2750 after 120 T/RH cycles. The bending strength experienced similar changes with treatment and cycling: from 3.2 to 9.1 MPa with ethyl silicate treatment and decreasing to 6.3 MPa with cycling. With forty days of immersion in water, the ultrasonic velocity fell from 3750 m/sec. to 2800 m/sec. (a drop similar to samples subjected to temperature and humidity cycling), while untreated granite showed little or no change with the same immersion. This indicates that it is the initial positive effects of the treatment that is being degraded by cycling and immersion and not the properties of the stone itself.

All of the evaluations cited so far, either field or laboratory, have been for stone exposed outdoors or for samples exposed to laboratory conditions simulating outdoor conditions. Thickett, Lee, and Bradley (2000) offer a rare look into the service life of alkoxysilane consolidants in a museum environment. These consolidants were applied to ancient Egyptian limestone sculptures, many contaminated with salts that engendered deterioration of the stone. In 1995 they examined a block statue treated with Wacker OH ten years earlier and found that "the treatment had not halted decay" (p. 506). X-ray diffraction performed on samples removed from powdering areas confirmed that sodium chloride was present and "caus[ed] damage to the surface fabric of the stone" (p. 506). They also evaluated samples of the gel removed from the sculpture. For samples removed in 1999, scanning electron microscopy demonstrated a breakdown in the gel network when compared to samples removed and examined in 1985. Other sculptures with lower salt contents, which were treated with Wacker OH in 1984, were still stable after fifteen years.

It is now time to bring together the information from the laboratory and the field and interpret and summarize what it says about the service life of alkoxysilane consolidation treatments on stone. Let us look first at the lessons from the laboratory.

- Mechanical testing of samples treated in the laboratory can give the upper limit of what is attainable under field conditions (Wendler et al. 1992). It can also determine the quality of a treatment performed in the field.
- When comparing the performance of consolidants in mechanical testing, it is important to look at both the relative strength increases and the absolute strength of treated samples. The relative strength increases in silicate rocks such as sandstones are usually larger than those in carbonate rocks such as limestones. The absolute strengths may, however, be similar.
- Artificial aging and experiments that attempt to simulate field conditions such as wet-dry and freeze-thaw cycling have sometimes brought about significant losses in mechanical properties in only a few cycles. These results underscore the need to make frequent site visits to determine when retreatment is necessary. The environment and the stone (condition and type) can also work against stabilization with alkoxysilanes such that it may be inadvisable to carry out the treatment.

There are equally important lessons from the field.

- It is difficult and unwise to predict or categorize the performance of alkoxysilane consolidants based on simplistic designations of stone type, for example, carbonate versus silicate or limestone versus sandstone.

- It is essential to determine the conditions present in and on the stone before prescribing consolidation with alkoxysilanes. Conditions such as scaling and flaking are often not stabilized by consolidation with alkoxysilanes. This is not a failure of the treatment but an inappropriate use. If the gaps between the stone substrate and the scales and flakes are small enough, such as for the sandstone from Bethesda Terrace (see fig. 6.10), and the mineralogy conducive to cohesion (quartz and feldspars), then the treatment can be successful and last up to twenty years.

- The structure of the stone can be equally important to the success of the treatment. Alkoxysilane consolidants can reduce granular disintegration in both silicate and carbonate rocks if the size and geometry of the intergranular spaces allow. In the examples noted above, the dolomitic limestones at both Howden Minster and the Fuentidueña Apse at the Cloisters exhibited granular disintegration that was poorly stabilized by alkoxysilane consolidants because the intergranular spaces are too large and too pocketlike (see fig. 6.13).

- It can be difficult to control the quality of consolidation treatments in the field. This lack of control can lead to overconsolidation, inadequate strengthening, or abrupt changes in the depth-strength profile. In fact, overconsolidated surfaces on stones that are susceptible to hygric dilatation can lead to scaling and flaking. In this context it may be wise to reconsider Bailey and Schaffer's (1964:n.p.) negative assessment of ethyl silicate treatments: "On stone of relatively poor quality its [ethyl silicate's] use has been followed by scaling of the treated surfaces." The performance of a conservation material cannot be adequately assessed without first knowing that it has been properly applied.

In Oliver's (2002:43) survey of treatments at a National Park Service site in the western United States, she states, "The effectiveness of

Figure 6.13

The porous, coarse-grained, and heterogeneous dolomitic limestones from Howden Minster (left) and the Fuentidueña Apse (right) have pore spaces too large (0.15–0.3 mm) and too pocketlike to be properly filled and bridged with alkoxysilane gels.

the ethyl silicate treatments depended on three principal factors: the original composition of the stone, its present condition, and, most importantly, the extent to which the causes of deterioration could be removed." This quotation serves as both the leitmotif and a succinct summary of this chapter. It is tempting to fall back on the old adage, nothing lasts forever. While this is undoubtedly true for alkoxysilane consolidants, it is also quite imprecise. I hope this discussion has provided a better and more precise understanding of some of the conditions that limit the service life of alkoxysilane consolidants for stone.

Notes

1. English Heritage is the United Kingdom's statutory adviser on the historic environment. Officially known as the Historic Buildings and Monuments Commission for England, English Heritage is an Executive Nondepartmental Public Body sponsored by the Department for Culture, Media and Sport (DCMS). Its powers and responsibilities are set out in the 1983 National Heritage Act.

2. As explained in chapter 4, the consolidating agent in Brethane is methyltrimethoxysilane, and the MTMOS is mixed with ethanol, water, and a catalyst immediately before use, usually in the following proportions: 64 ml MTMOS, 22 ml industrial methylated spirits, 13 ml water, and 1 ml of a 6% v/v solution of Manosec Lead 36 in dry white spirits. Proportions are sometimes varied depending on atmospheric temperature. Industrial methylated spirits, or IMS, is a form of denatured alcohol. Manosec Lead 36, produced by Manchem Ltd., is based on lead naphthenate and has often been used as a paint drier. White spirits is a form of mineral spirits.

3. According to Dimes (Ashurst and Dimes 1990), "clunch" is the term commonly used by masons for chalk—the soft, earthy, fine-textured limestone of marine origin. He also indicates the term should be restricted to the types of chalk found in East Anglia.

4. Biosparite is defined as a limestone with more than 10% allochems and a dominantly sparite matrix (Tomkeieff 1983).

5. Fort, Azcona, and Mingarro (2000) indicated that consolidated stone resoiled more rapidly than untreated areas.

6. As noted in chapter 5, alkoxysilane consolidants can suppress or deter biological growth by drying out the stone and by the toxic effects of some catalysts such as lead naphthenate in Brethane and dibutyltindilaurate in many TEOS-based consolidants. On the other hand, some consolidants can promote biological growth. See Koestler 2000; Koestler et al. 1986.

7. Sometimes referred to as magnesian limestone. Tomkeieff (1983) offers the following definition: (1) a limestone containing about 5 to 15% magnesium carbonate but in which dolomite cannot be detected, (2) a term also used loosely for a dolomitic limestone, (3) a stratigraphical division of the Permian rocks in England.

8. This stone contains the mineral dolomite, which breaks down in acid rain or dry deposition in the following way:

$$(Ca_{0.5}Mg_{0.5})CO_3 + H_2SO_4 \rightarrow CaSO_4 \bullet 2H_2O + MgSO_4 \bullet 6 \text{ or } 7H_2O$$

(i.e., hexahydrite or epsomite salts).

9. The salts were analyzed by x-ray diffraction.

10. My field notes from 1996 indicate that both the treated and the untreated dolomite at Howden Minster have a moderately high level of granular disintegration.

11. Including the two low values does nothing to change the relative averages of porosity for limestones versus sandstones.

12. Efflorescences can increase after treatment because consolidation can stabilize stone that might otherwise be lost during the movement of moisture and the salts it carries. In this scenario the increase in efflorescence is a sign that the treatment is actually working in its function as a consolidant.

13. Bethesda Fountain and Terrace were designed by Charles Vaux; the sculpture was executed by Emma Stebbins. See Willensky and White's *AIA Guide to New York City* (1988).

14. The stone was analyzed by x-ray diffraction and found to comprise primarily quartz with smaller amounts of albite feldspar and even smaller amounts of illite and chlorite. X-ray fluorescence found about 6% iron. The stone is said to be from the Wallace quarries of Nova Scotia and is green when freshly quarried. It weathers to a tan color, probably due to changes in the hydration of iron oxides.

15. This designation was provided by the geologist Lorenzo Lazzarini. See Charola et al. 1986.

16. Among the salts identified by Charola was humberstonite, found naturally only in the Atacama Desert in Chile.

17. An important unanswered and perhaps unanswerable question is what happened to the stone between 1993 and 1996 that caused a fundamentally different ability to resist forces of deterioration that had been present for centuries?

18. Although not explicitly stated, the consolidant was probably *Conservare* OH.

19. It is not clear what Oliver means by this statement. It may be that she believes that calcite is a more stable cementing material than clay or that there is less calcite cement such that grains of quartz and feldspar are more closely in contact with each other.

20. "Friable" means easily crumbled and, as a condition of stone, is sometimes confused with granular disintegration. Stone that is truly friable also usually exhibits granular disintegration. It can be argued that a stone is friable if it exhibits granular disintegration to a depth greater than one or two millimeters.

21. For a more complete rendering and rationale of this treatment, see Matero and Oliver 1997.

22. Schilfsandstein is a clay-rich sandstone susceptible to contour scaling by hygric dilatation (Snethlage and Wendler 1991).

23. Stadlbauer et al. (1996) give the following description of Baumberg sandstone: 10 volume % visible pore space, 60 volume % fine-grained components (predominantly calcareous detritus, quartz, and glauconite) surrounded by a carbonate matrix (about 30 volume %) containing quartz, clay minerals, and other silicates. It contains less than 50 volume % quartz and therefore would more accurately be called a sandy limestone.

24. Their outdoor weathering consisted of an unmonitored two-year exposure on the roof of the Old Treasury Building in Melbourne, Australia, and the artificial weathering consisted of 2000 hours of exposure in cycles of 4 hours of UVB radiation at 70°C followed by 4 hours of condensation at 50°C.

25. The disadvantage of this test is that it assesses only the hardness of the surface. This test may give only a partial, or even incorrect, assessment of the quality of the consolidation treatment in depth (Sattler and Snethlage 1988).

26. Their protocol consisted of the following: dry samples at 105°C for 24 hours, immersion in water for 24 hours under reduced pressure, cycling in a cold chamber from +10 °C to −10 °C one cycle per day for 10 days.

27. Unpublished manuscript on the assessment of conditions at Victoria Mansion, Portland, Maine, by Building Conservation Associates, New York.

28. Portland brownstone is a clay-bearing sandstone quarried in Connecticut and frequently used for construction in the northeastern United States in the last half of the nineteenth century. The elastic moduli were measured using ultra-sonic velocity pulses, and data generated by this technique for clay-bearing stones may be suspect.

Chapter 7
Recent Developments and Final Thoughts

In making the journey defined by the previous six chapters, you will understand that the words "final thoughts" apply only to the pages before you and not the subject of alkoxysilanes and the consolidation of stone. The story of the development of alkoxysilane treatments and the understanding of their use on stone will continue. This book attempts to summarize the subject to date and to emphasize, if only by inference, that this development and understanding are a process of constant accrual and reevaluation. It would be difficult to argue otherwise given that alkoxysilanes were first suggested as stone consolidants in 1861.

The introduction of a new conservation material is often accompanied by great optimism and enthusiasm on the part of both the inventor and the first users. The excitement is equally often a signal of how great is the need to be filled. Alkoxysilanes are a promising material. They are not, however, the panacea for the problems of stone consolidation but one tool in a set to address these problems. The use of such a tool requires that its strengths, weaknesses, and limitations are understood and that it be used under and for the right conditions. We are still in the process of gaining that understanding.

Somewhat surprisingly, formulations have changed little over the past thirty years, and the partially polymerized tetraethoxysilane (ethyl silicate) that Hoffmann (1861) suggested is still the basis of most commercial products today. In the past ten years some modifications in formulations and treatments have grown out of an understanding of some of the limitations of alkoxysilanes as stone consolidants. Two generally recognized drawbacks of these consolidants are the inability to bond to calcite and the tendency for gels to crack during shrinkage and drying due to their brittleness.

Weiss, Slavid, and Wheeler (2002) addressed the problem of bonding to calcite by chemically altering the mineral's surface. Calcite by itself contains few hydroxyl groups for alkoxysilanes to condense with. They designed a treatment to create a hydroxyl-rich surface by reacting calcite with ammonium hydrogen tartrate in a pH-balanced solution known commercially as HCT, or hydroxy conversion treatment.[1] As can be seen in figure 7.1, ammonium tartrate reacts with calcite to form calcium tartrate. (It should be made clear from the start that some calcite is consumed in the process.)[2] The SEMs in figures 7.2a and 7.2b demonstrate

Figure 7.1

In the HCT treatment, a pH-balanced solution of ammonium hydrogen tartrate reacts
with a calcite substrate. Some calcite is consumed in the reaction, which also leaves
calcium tartrate attached to the surface.
The bonded tartrate carries two hydroxyl
groups that can react with alkoxysilanes.

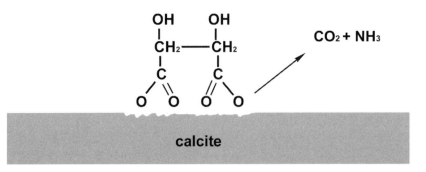

Figure 7.2a (left)

This scanning electron micrograph of an
HCT-treated surface of Iceland spar
demonstrates that a single treatment of
HCT does not completely cover the surface of calcite with crystals of calcium
tartrate. Photo: Mark Wypyski

Figure 7.2b (right)

A second treatment with HCT has completely converted the surface to calcium
tartrate. Photo: Mark Wypyski

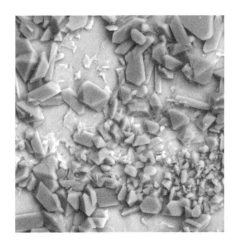

that at least two treatments are needed to completely convert the surface
of a crystal of calcite to calcium tartrate.

The newly formed calcium tartrate can condense with alkoxysilanes to create Si-O-C linkages that bind the gel to the calcite (fig. 7.3).
The hydrolytic stability of these Si-O-C bonds is not known. In theory
they can be hydrolyzed and therewith separate the gel from the mineral
substrate. As discussed in chapter 2, not all Si-O-C bonds are equal: the
larger the organic group attached to the "C," the greater the steric hindrance that resists hydrolytic cleavage of the bond. The tartrate group is
large and itself hindered by its attachment to the calcite substrate. These
factors would make hydrolysis difficult.[3]

Field testing of HCT is still rather limited, but laboratory tests
have shown some promising results. For example, cleaved pieces of
Iceland spar[4] that have been treated with HCT followed by *Conservare*

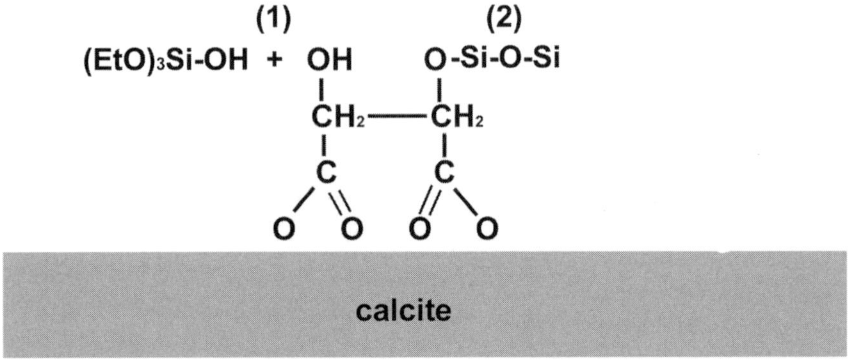

Figure 7.3

In step (1) ethyl silicate reacts with the OH
groups on calcium tartrate. This condensation reaction creates Si-O-C linkages
between the consolidant and the substrate.

Figure 7.4

The pieces of silica gel visible in this photograph are fully attached to the calcite substrate due to an initial treatment with HCT. Ethyl silicate treatments on calcite without HCT deposit gel pieces that can be dislodged with the slightest touch.

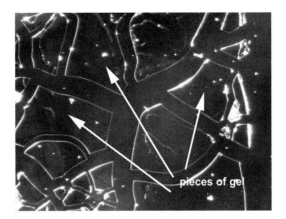

OH100 create a monolithic "sandwich" that remains intact even after several months of immersion in water, and the gel remains adhered to the Iceland spar even after the "sandwich" is split open (fig. 7.4). Similar pieces of spar treated with OH100 do not form monoliths, nor is the gel attached to the substrate.

Another approach that has been advanced to create linkages across the interface between calcite and alkoxysilane-derived gels is to employ alkoxysilane coupling agents. These coupling agents are compounds that have hydrolyzable alkoxy groups (usually three) attached to the central silicon atom and a fourth group attached to silicon with an unhydrolyzable Si-C bond. Coupling agents are fundamental to improving the strength of composite materials that comprise inorganic fillers surrounded by organic resins. In the fabrication of a composite material made, for example, of glass fibers and epoxy resin, the two different functional groups on the alkoxysilane coupling agent offer separate compatibility with the fibers and the resin. Alkoxy groups provide silanols to link with similar groups on the glass fibers and the organic, or R, group links with the epoxy resin (fig. 7.5). The resulting material is much stronger than its equivalent without the coupling agent (Plueddemann 1991). Wheeler et al. (1991) and Wheeler et al. (2000) used a similar approach to improve the performance of alkoxysilane consolidants on marble and limestone. In this case, the inorganic substrate or "filler" is the stone (calcite), and the resin is the alkoxysilane consolidant.[5] They reasoned that R groups could be found that were compatible with, or

Figure 7.5

Silanols on the glass fiber (1) condense with silanols on the coupling agent (2). The coupling agent has an R group compatible with the epoxy resin (3). The two different functional groups on the coupling agent create linkages across the interface of the fiber and the resin.

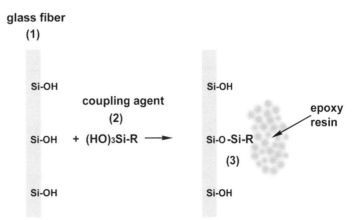

Figure 7.6

The treatment rationale presented here is similar to that shown in figure 7.5. Here, the R group on the coupling agent (2) is chosen to be compatible with the calcite substrate (1). The silanol or silicate groups on the coupling agent are compatible with ethyl silicate or the derived silica gel (3).

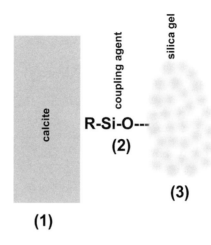

bond to, calcite and that the hydrolyzed alkoxy groups on the coupling agent would form silanols to condense with the alkoxysilane consolidant (fig. 7.6). Coupling agents have been tested both as primers followed by TEOS-based consolidants and as integral blends in these same consolidants. Marked improvements in modulus of rupture—up to 153% over and above treatments without coupling agents—were noted. Field testing has yet been performed on these materials, and they have not been incorporated into commercial products.

The second major problem associated with alkoxysilane consolidants is their brittleness and the related tendency to crack with drying and shrinkage (see chap. 4). Wendler (1996) and Snethlage and Wendler (1991) chose to address this problem by making the gel itself more compliant (or less brittle) by introducing segments of linear siloxanes that linked up with the larger silicate networks of the gel (fig. 7.7). Both the gel and the consolidated stone are less brittle, and there is a less abrupt transition between deteriorated stone that has been consolidated and undeteriorated or unconsolidated stone that lies below the surface. This smoother transition leaves treated stone less susceptible to contour scaling. These ideas have been incorporated into a line of Remmers products

Figure 7.7

Stress-strain plot for treated and untreated sandstone: with many ethyl silicate consolidants (SAE = silicic acid ethyl esters), an increase in both strength and elastic modulus is observed. The resulting brittleness may make the stone more susceptible to damage. Good strength increases can be attained with lower elastic moduli by introducing elastic bridges into the TEOS oligomers of the consolidant (SAE elastified). The modulus is obtained from the slope of the stress-strain curve. The slope of the SAE elastified treated stone is clearly less than the unelastified stone (SAE OH) (E. Wendler 1996).

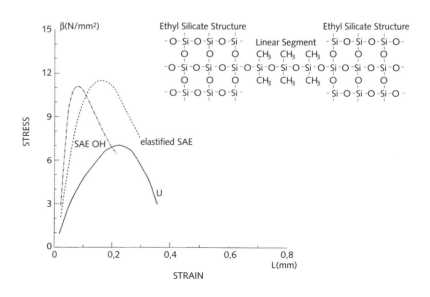

that has had several years of field testing.[6] We look forward to seeing more field reports and laboratory testing on these products.

Escalante, Valenza, and Scherer (2000) took a different approach to gel shrinkage and cracking. They noted that drying shrinkage is controlled by competition between the capillary pressure that drives contraction and the elastic modulus of the gel that resists contraction (Smith, Scherer, and Anderson 1995). Their approach involved adding small particles (~ 2 μm) in silica sols. The addition of the particles increased the moduli and pore sizes, and the resulting gels experience reduced shrinkage and little cracking.[7] Scherer has continued to work on the development of these systems. Although no field testing has taken place to date, they seem to hold some promise for consolidating highly porous stone.[8]

The importance of laboratory and field testing (including periodic reevaluations) has been stressed throughout this book. As this book comes to a close it is worthwhile to comment on some of the laboratory tests and, in a more general way, on field evaluations. The comments on field evaluations can be dispensed with very quickly: more on the order of Martin et al. (2002) are needed. In particular, the only substantive field evaluation of alkoxysilane-treated marble that has involved returning to the site to determine how well the treatment has performed is Roby (1996). The increased use of these consolidants on marble make the need for further work all the more pressing.

The subject of laboratory tests is larger and more complex and can be touched on only briefly here. Testing usually involves determining mechanical properties before and after treatment. Of the many tests performed, compressive strength measurements are probably the least useful. Stones generally have rather high compressive strengths, and while they may be increased by consolidation treatments, these strength increases are not very useful in resisting the larger forces that produce compression failure. Other strength tests—tensile, flexure (3- and 4-point bend, biaxial ring, etc.), axial split—appear to give more useful information because so often the actual mode of failure is in response to small- or medium-scale tensile stresses. Still, questions remain: Which is the best among these tests? How do results from one test relate to the others? What are the fundamental limitations of these tests specifically related to testing consolidated stone? The question of limitations (and anomalies) arises with the technique of ultrasonic velocity (USV) measurements used to determine another bulk property, elastic modulus. In the case of marble, the technique is highly sensitive to moisture content: the USV rises dramatically when a few drops of water are added to small samples. It has been reported that clay-bearing stones can also give anomalous results (G. Scherer, pers. comm., 2004).

In recent years tests have been developed that try to relate more closely to the form or manifestation of deterioration that the stone exhibits, that is, the condition that one seeks to mend with consolidant. For example, microabrasion testing introduced by Phillips (1982) and refined by Ginell et al. (1991) has been used to evaluate the ability of consolidants to stabilize stone against losses due to granular disintegration. The usefulness of this test is borne out by the fact that the measurement of many of the bulk properties mentioned above often show little difference between treated and untreated marble, but the reduction in losses brought about by consolidation and measured by microabrasion are significant. Nonetheless, the test, as it is currently configured by the above-mentioned authors, has its limitations. If the grain size of the marble is larger than approximately two millimeters, then the test is almost useless.[9] Finally, drill resistance measurements, which can also be performed in the field as well as the laboratory, may offer a relevant way of evaluating treatments on significantly weakened stone. The technique also yields depth-strength profiles that are useful in determining the quality of a treatment: How well has the treatment been executed? Has the consolidant reached adequate depth? Is the surface overconsolidated? Is there an abrupt change in strength with depth?

The issue of quality of treatment cannot be stressed strongly enough. A poorly executed treatment employing the ideal material usually leads to failure and possibly rejection of the treatment material. Scientists may be able to determine ex post facto that a treatment was not successfully carried out, and these same scientists may be able to perform tests before treatment that provide data helpful to conservators, but the success of the treatment lies, for the most part, in the hands, the eyes, and the other senses of the conservators. Consolidation of stone with alkoxysilanes, like all conservation treatments, involves elements of the cognitive, the intuitive, and the sensory. Scientists tend to favor the cognitive—the so-called objective form of knowing—over the more subjective but equally important intuitive and sensory forms of knowing that make up so much of the world of the conservator. Our goal should be a dynamic fusion of these ways of knowing and not their separation. The reward will be better conservation of our heritage.

Notes

1. HCT is sold by ProSoCo, Lawrence, Kansas (www.prosoco.com).

2. Each application of HCT would be the equivalent of one acid rain event.

3. Peeler (1959) indicated that di-*t*-butoxy-di-octoxysilane is so unreactive with water that it can be used as a high-temperature hydraulic fluid in aircraft engines.

4. Iceland spar is a very pure and transparent form of calcite.

5. Aminoorganosilane was suggested by Cormerois (1978) to improve the film-forming properties of consolidants. Others who have tested coupling agents are Munnikendam (1971), Nagy et al. (1997), and Rao et al. (1996).

6. These products, along with their elastomeric nonhydrophobic counterparts, are the following Remmers products: Funcosil 300 E and STE and Funcosil 500 E and STE. According to the 20 October 1998 product literature, they also contain ethanol. Boos et al. (1996) tested elastified and unelastified versions of the same product and found somewhat lower elastic moduli for two German sandstones (Bentheim and Buch); strength increases were about the same for the two different treatments.

7. Scherer and Wheeler (1997) have shown that unmodified gels had low permeability and small pores.

8. The particle-modified solutions are higher in viscosity and may not be able to enter small pores.

9. Microabrasion testing involves directing a pencil-tip-sized nozzle that delivers a fine powder of abrasive at the stone sample. If the grain sizes of the stone are large, the abrasive creates holes in individual grains instead of dislodging them.

References

Aelion, R., A. Loebel, and F. Eirich. 1950. Hydrolysis of ethyl silicate. *Journal of the American Chemical Society* 72:5705–12.

Aguzzi, F., A. Fiumara, A. Peroni, R. Ponci, V. Riganti, R. Rossetti, F. Soggetti, and F. Veniale. 1973. L'arenaria della Basilica di S. Michele in Pavia: Ricerche sull'alterazione e sugli effetti dei trattementi conservativi. *Atti Societa Italiana di Scienze Naturali* 114 (4):403–64.

Alaimo, R., R. Giarrusso, L. Lazzarini, F. Mannuccia, and P. Meli. 1996. The conservation problems of the Theatre of Eraclea Minoa (Sicily). In *Eighth International Congress on Deterioration and Conservation of Stone*, ed. J. Riederer, 1085–95. Berlin: Möller Druck und Verlag.

Alessandrini, G., R. Bonecchi, E. Broglia, R. Bugini, R. Negrotti, R. Peruzzi, L. Toniolo, and L. Formica. 1988. Les colonnes de San Lorenzo (Milan, Italie): Identification des matériaux, causes d'alteration, conservation. In *The Engineering Geology of Ancient Works, Monuments and Historic Sites*, vol. 2, ed. P. Marinos and G. Koukis, 925–32. 4 vols. Rotterdam: A. A. Balkema.

Anon. 1972. Building preservatives fall on stony ground. *New Scientist*, 31 August, 437.

Anon. 1861. Stone-preserving processes: Royal Institute of British Architects. *The Builder* 19: 103–5.

Arkles, B., ed. 1998. *Gelest Catalog.* Tullytown, Pa.: Gelest.

Arnold, L. 1978. The preservation of stone by impregnation with silanes. *Newsletter of the Council for Places of Worship* 24:4–6.

Arnold, L., D. Honeyborne, and C. A. Price. 1976. Conservation of natural stone. *Chemistry and Industry* (April):345–47.

Arnold, L., and C. A. Price. 1976. The laboratory assessment of stone preservatives. In *The Conservation of Stone I*, ed. R. Rossi-Manaresi, 695–704. Bologna: Centro per la Conservazione delle Sculture all'Aperto.

Ashurst, J., and F. G. Dimes. 1990. *Conservation of Building and Decorative Stone.* Vol. 2. Boston: Butterworth-Heinemann.

ASTM C97-90. 1993. *Standard Test Method for Absorption and Bulk Specific Gravity of Dimension Stone.* Philadelphia: American Society of Testing and Material.

ASTM D789-98. 1998. *Standard Test Methods for Determination of Relative Viscosity and Moisture Content of Polyamide.* Philadelphia: American Society of Testing and Material.

ASTM D4878-98. 1998. *Standard Test Methods for Polyurethane Raw Materials Determination of Viscosity of Polyols.* Philadelphia: American Society of Testing and Material.

Bailey, T. A., and R. J. Schaffer. 1964. Report on stone preservation experiments. Ancient Monuments Branch, Ministry of Public Works, U.K., November.

Bates, R., and J. A. Jackson. 1984. *Dictionary of Geological Terms.* 3d ed. Garden City, N.Y.: Anchor-Doubleday.

Bensaude-Vincent, B., and I. Stengers. 1996. *A History of Chemistry.* Cambridge, Mass.: Harvard University Press.

Berry, J., and C. Price. 1989. The movement of salts in consolidated stone. In *The Conservation of Monuments in the Mediterranean Basin*, ed. F. Zezza, 845–48. Bari: Grafo Edizioni.

Berti, P. 1979. Prove di laboratorio su elementi di mattoni trattati con resine siliconiche. In *Il Mattone di Venezia*, 439–46. Venice: Laboratorio per lo Studio della Dinamica delle Grandi Masse del CNR e dell'Universita di Venezia.

Berzelius, J. J. 1824. Untersuchungen über die Flusspathsäure und deren merkwürdigsten Verbindungen. *Annalen der Physik und Chemie* 1:169–230.

Blasej, J., J. Doubrava, and J. Rathousky. 1959. Pousiti organokremicitych latek pro konservaci a resauraci casti piscovcoveho zabradli letohradku v kralovske zahrade. *Zpravy Pamatkove Pece* 19:69–80.

Boos, M., J. Grobe, G. Hilbert, and J. Müller-Rochholz. 1996. Modified elastic silicic-acid ester applied on natural stone and tests of their efficiency. In *Eighth International Congress on Deterioration and Conservation of Stone*, ed. J. Riederer, 1179–85. Berlin: Möller Druck und Verlag.

Bosch, E. 1972. Use of silicones in conservation of monuments. In *First International Symposium on Deterioration of Building Stones*, 11–16 September. Chambery: Imprimeries Reunies, 1973, 21-26.

Bosch, E., M. Roth, and K. Gogolok. 1973. Binder composition for inorganic Compounds. German Patent Application 2,318,494. 12 April.

Bosch, E., M. Roth, and K. Gogolok. 1976. Binder composition for inorganic Compounds. U.S. Patent Application 3,995,988. 11 May.

Boyle, R. S. 1980. *Chemistry and Technology of Lime and Limestone.* 2d ed. New York: Wiley.

Bradley, S. M. 1985. Evaluation of organosilanes for use in consolidation of sculpture displayed indoors. In *Fifth International Congress on Deterioration and Conservation of Stone*, ed. G. Felix, 759–67. Lausanne: Presses Polytechniques Romandes.

Bradley, S. M., and S. B. Hanna. 1986. The effect of soluble salt movement on the conservation of an Egyptian limestone standing figure. In *Case Studies in the Conservation of Stone and Wall Paintings*, ed. N. S. Brommelle and P. Smith, 57-61. London: IIC.

Brinker, C. J., and G. Scherer. 1990. *Sol-Gel Science: The Physics and Chemistry of Sol-Gel Processing.* London: Academic Press.

Brus, J., and P. Kotlik. 1996. Cracking of organosilicone stone consolidants in gel form. *Studies in Conservation* 41 (1):55–59.

Butlin, R. N., T. J. S. Yates, and W. Martin. 1995. Comparison of traditional and modern treatments for conserving stone. In *Methods of Evaluating Products for the Conservation of Porous Building Materials in Monuments,* ed. M. L. Tabasso, 111–19. Rome: ICCROM.

Caselli, A., and D. Kagi. 1995. Methods used to evaluate the efficacy of consolidants on an Australian sandstone. In *Methods of Evaluating Products for the Conservation of Porous Building Materials in Monuments,* ed. M. L. Tabasso, 121–30. Rome: ICCROM.

Charola, A. E., and R. J. Koestler. 1986. Scanning electron microscopy in the evaluation of consolidation treatments for stone. In *Scanning Electron Microscopy,* vol. 2, ed. R. P. Becker and G. M. Roomans, 479–84. Chicago: Scanning Electron Microscopy.

Charola, A. E., R. J. Koestler, and G. Lombardi, eds. 1994. *Lavas and Volcanic Tuffs.* Rome: ICCROM.

Charola, A. E., L. Lazzarini, G. Wheeler, and R. J. Koestler. 1986. The Spanish apse from San Martin de Fuentidueña at the Cloisters, Metropolitan Museum of Art, New York. In *Case Studies in the Conservation of Stone and Wall Paintings,* ed. N. S. Brommelle and P. Smith, 18–21. London: IIC.

Charola, A. E., R. Rossi-Manaresi, R. J. Koestler, G. Wheeler, and A. Tucci. 1984. SEM examination of limestones treated with silanes or prepolymerized silicone resin in solution. In *Adhesives and Consolidants,* ed. N. S. Brommelle, E. M. Pye, P. Smith, and G. Thomson, 182–84. London: IIC.

Charola, A. E., G. Wheeler, and G. G. Freund. 1984. The influence of relative humidity in the polymerization of methyltrimethoxysilane. In *Adhesives and Consolidants,* ed. N. S. Brommelle, E. M. Pye, P. Smith, and G. Thomson, 177–81. London: IIC.

Charola, A. E., G. Wheeler, and R. J. Koestler. 1983. Treatment of the Abydos reliefs: Preliminary investigations. *Fourth International Congress on Deterioration and Preservation of Stone Objects,* ed. K. L. Gauri and J. A. Gwinn, 77–88. Louisville: University of Louisville.

Chiari, G. 1987. Consolidation of adobe with ethyl silicate: Control of long-term effects using SEM. In *Fifth International Meeting of Experts on the Conservation of Earthen Architecture,* 25–32. Rome: ICCROM-CRATerre.

Chiari, G. 1980. Treatment of adobe friezes in Peru. In *Third International Symposium on Mud-Brick Preservation,* 39–45. Ankara: ICOM-ICOMOS.

Clarke, B. L., and J. Ashurst. 1972. *Stone Preservation Experiments.* London: H.M. Stationery Office.

Cogan, H. D., and C. A. Setterstrom. 1946. Properties of ethyl silicate. *Chemical and Engineering News* 24 (18):2499–2501.

Cogan, H. D., and C. A. Setterstrom. 1947. Ethyl silicates. *Industrial and Engineering Chemistry* 39 (11):1364–68.

Cormerois, P. 1978. Traitements préventif et curatif des structures. In *Preprints for the International Symposium: Deterioration and Protection of Stone Monuments, 5–9 June 1978,* Volume 2, Number 6.5, Paris: RILEM and UNESCO.

Costa, D., and J. Delgado Rodrigues. 1996. Assessment of colour changes due to treatment products in heterochromatic stones. In *Conservation of Granitic Rocks,* ed. J. Delgado Rodrigues and D. Costa, 95–101. Lisbon: LNEC.

Danehey, C., G. Wheeler, and H. S. Su. 1992. The influence of quartz and calcite on the polymerization of methyltrimethoxysilane. In *Seventh International Congress on Deterioration and Conservation of Stone,* ed. J. Delgado Rodrigues, F. Henriques, and F. Telmo Jeremias, 1043–52. Lisbon: LNEC.

Debsikdar, J. C. 1986. Effect of the nature of the sol-gel transition on the oxide content and microstructure of silica gel. *Advanced Ceramic Materials* 1 (1):93–98.

De Witte, E., and K. Bos. 1992. Conservation of ferruginous sandstone used in northern Belgium. In *Seventh International Congress on Deterioration and Conservation of Stone,* ed. J. Delgado Rodrigues, F. Henriques, and F. Telmo Jeremias, 1113–21. Lisbon: LNEC.

De Witte, E., A. E. Charola, and R. P. Sherryl. 1985. Preliminary tests on commercial stone consolidants. In *Fifth International Congress on Deterioration and Conservation of Stone,* ed. G. Felix and V. Furlan, 709–18. Lausanne: Presses Polytechniques Romandes.

Dinsmore, J. 1987. Consideration of adhesion in the use of silane-based consolidants. *Conservator* 11:26–29.

Domaslowski, W. 1987–88. The mechanism of polymer migration in porous stones. *Wiener Berichte über Naturwissenschaft in der Kunst* 4–5:402–25.

Domaslowski, W., and J. W. Lukaszewicz. 1988. Possibilities of silica application in consolidation of stone monuments. In *Sixth International Congress on Deterioration and Conservation of Stone,* ed. J. Ciabach, 563–76. Torun: Nicholas Copernicus University.

Driers and Drying. 1981. Manchester, UK: Manchem Ltd.

Duran-Suarez, A., J. Garcia-Beltran, and J. Rodriguez-Gordillo. 1995. Colorimetric cataloguing of stone materials (biocalcarenite) and evaluation of the chromatic effects of different restoring materials. *Science of Total Environment* 167:171–80.

Eaborn, C. 1960. *Organosilicon Compounds.* New York: Academic Press.

Ebelmen, J. J. von. 1846. Untersuchungen über die Verbindungen der Borsäure mit Aether. *Annalen der Chemie und Pharmacie* 57:319–53.

Elfving, P., and U. Jäglid. 1992. Silane bonding to various mineral surfaces. Report OOK 92:01, ISSN 0283-8575. Götenborg, Sweden: Department of Inorganic Chemistry, Chalmers University of Technology.

Ellis, C. 1935. Inorganic resins. In *The Chemistry of Synthetic Resins,* vol. 2, 1235–39. New York: Reinhold.

Emblem, H. G. 1947. Organo-silicon compounds in paint media: A review of recent developments. *Paint Manufacture* 17 (7):239–40.

Emblem, H. G. 1948a. Recent developments in silicon ester paints. *Paint Manufacture* 18:359–60.

Emblem, H. G. 1948b. Silicon ester paints. *Paint Technology* 13 (2):309–11.

Escalante, M., J. Valenza, and G. Scherer. 2000. Compatible consolidants from particle-modified gels. In *Ninth International Congress on Deterioration and Conservation of Stone*, vol. 2, ed. V. Fassina, 459–65. Amsterdam: Elsevier.

Ettl, H., and H. Schuh. 1989. Konservierende Festigung von Sandsteinen mit Kieselsäureethylester. *Bautenschutz und Bausanierung* 12:35–38.

Falcone, J. S., ed. 1982. *Soluble Silicates*. Washington, D.C.: American Chemical Society.

Felix, C. 1995. Peut-on consolider les grès tendres du Plateau Suisse avec le silicate d'ethyle? In *Preservation and Restoration of Cultural Heritage*, ed. R. Pancella, 267–74. Lausanne: Ecole Polytechnique Federale de Lausanne.

Felix, C., and V. Furlan. 1994. Variations dimensionnelles de grès et calcaires, liees a leur consolidation avec un silicate d'ethylene. In *Third International Symposium on the Conservation of Monuments in the Mediterranean Basin*, ed. V. Fassina, H. Ott, and F. Zezza, 855–59. Venice: Soprintendenza ai Beni Artisitici e Storici di Venezia.

Fitzner, B. 1988. Porosity properties of naturally or artificially weathered sandstones. In *Sixth International Congress on Deterioration and Conservation of Stone*, ed. J. Ciabach, 236–45. Torun: Nicholas Copernicus University.

Fitzner, B., K. Heinrichs, and R. Kownatzki. 1995. Weathering forms—classification and mapping (Verwitterungsformen—Klassifizierung und Kartierung). In *Denkmalpflege und Naturwissenschaft: Natursteinkonservierung I*, ed. R. Snethlage, M. Sandfort, and H.-U. Schirmer, 41–88. Berlin: Ernst & Sohn.

Fitzner, B., and R. Snethlage. 1982. Einfluss der porenradienvertielung auf das verwitterungsverhalten ausgewählter sandsteine. *Bautenschutz und Bausanierung* 5:97–103.

Fort, R., M. C. Lopez de Azcona, and F. Mingarro. 2000. Assessment of protective treatments based on their chromatic evolution: Limestone and granite in the Royal Museum of Madrid, Spain. In *Protection and Conservation of the Cultural Heritage of the Mediterranean Cities: Proceedings of the Fifth International Symposium on the Conservation of Monuments in the Mediterranean Basin, Sevilla, Spain, 5–8 April 2000*, ed. E. Galan and F. Zezza, 437–441. Lisses: Elsevier.

Fowkes, F. 1987. Role of acid-base interfacial bonding in adhesion. *Journal of Adhesion Science and Technology* 1 (1):7–27.

Frogner, P., and L. Sjoberg. Dissolution of tetraethylorthosilicate coatings on quartz grains in acid solution. In *Eighth International Congress on Deterioration and Conservation of Stone*, ed. J. Riederer, 1233–41. Berlin: Möller Druck und Verlag.

Galan, E., and I. Carretero. 1994. Estimation of the efficacy of conservation treatments applied to a permotriassic sandstone. In *Third International Symposium on the*

Conservation of Monuments in the Mediterranean Basin, ed. V. Fassina, H. Ott, and F. Zezza, 947–54. Venice: Soprintendenza ai Beni Artistici e Storici di Venezia.

Gardner, H. A., and G. G. Sward. 1932. Experiments to preserve fresco for exterior decoration. *American Paint and Varnish Manufacturer's Association Circular,* no. 421, 285–89.

General Electric Company. 1959. Improvements relating to organosilicon compounds. U. K. Patent 813,520. 21 May.

Ginell, W. S., R. Kumar, and E. Doehne. 1991. Conservation studies on limestone from the Maya site at Xunantunich, Belize. In *Materials Issues in Art and Archaeology IV,* ed. P. B. Vandiver, J. Druzik, J. L. G. Madrid, I. C. Freestone, and G. Wheeler, 813–21. Pittsburgh: Materials Research Society.

Goins, E. S. 1995. Alkoxysilane stone consolidants: The effect of the stone substrate on the polymerization process. Doctoral diss., University College London, University of London.

Goins, E. S., G. Wheeler, and S. A. Fleming, 1995. The influence of reaction parameters on the effectiveness of tetraethoxysilane based stone consolidants. In *Methods of Evaluating Products for the Conservation of Porous Building Materials in Monuments,* ed. M. L. Tabasso, 259–74. Rome: ICCROM.

Goins, E. S., G. Wheeler, D. Griffiths, and C. A. Price. 1996. The effect of sandstone, limestone, marble and sodium chloride on the polymerization of MTMOS solutions. In *Eighth International Congress on Deterioration and Conservation of Stone,* ed. J. Riederer, 1243–54. Berlin: Möller Druck und Verlag.

Graulich, W. 1933a. Kieselsaure-Ester als Lackfarben-Grundkorper. *Nitrocellulose* 4:61–62.

Graulich, W. 1933b. Zur Frage des Bautenschutzes: Neuzeitliche Steinkonservierung mit Kieselsaureester-Farben. *Tonindustrie-Zeitung und Keramische Rundschau* 57:677.

Grissom, C. 1996. Conservation of Neolithic lime plaster statues from 'Ain Ghazal. In *Archaeological Conservation and Its Consequences,* ed. A. Roy and P. Smith, 70–75. London: IIC.

Grissom, C. A., A. E. Charola, A. Boulton, and M. F. Mecklenburg. 1999. Evaluation over time of an ethyl silicate consolidant applied to ancient lime plaster. *Studies in Conservation* 44 (2):113–20.

Grissom, C. A., and N. R. Weiss. 1981. *Alkoxysilanes in the Conservation of Art and Architecture: 1861–1981.* In *Art and Archaeology Technical Abstracts* 18 (1), suppl.:150–202.

Hammecker, C., R. M. E. Alemany, and D. Jeannette. 1992. Geometry modifications of porous networks in carbonate rocks by ethyl silicate treatment. In *Seventh International Congress on Deterioration and Conservation of Stone,* ed. J. Delgado Rodrigues, F. Henriques, and F. Telmo Jeremias, 1053–62. Lisbon: LNEC.

Hanna, S. B. 1984. The use of organo-silanes for the treatment of limestone in an advanced state of deterioration. In *Adhesives and Consolidants,* ed. N. S. Brommelle, E. M. Pye, P. Smith, and G. Thomson, 171–76. London: IIC.

Hawley, G. G. 1981. *The Condensed Chemical Dictionary*. 10th ed. rev. New York: Van Nostrand Reinhold.

Heaton, N. 1930. The possibilities of inorganic paint vehicles. *Journal of the Oil and Colour Chemists' Association* 13:330–40.

Hempel, K. 1976. An improved method for vacuum consolidation of decayed stone sculpture. In *Second International Congress on Deterioration and Conservation of Stone*, ed. T. Skoulikidis, 163–66. Athens: Ministry of Culture and Science of Greece.

Hempel, K., and A. Moncrieff. 1972. Summary of work on marble conservation at the Victoria and Albert Museum Conservation Department up to August 1971. In *The Treatment of Stone*, ed. R. Rossi-Manaresi and G. Torraca, 165–81. Bologna: Centro per la Conservazione delle Sculture all'Aperto.

Hempel, K., and A. Moncrieff. 1976. Report on work since last meeting in Bologna, October 1971. In *The Conservation of Stone I*, ed. R. Rossi-Manaresi, 319–39. Bologna: Centro per la Conservazione delle Sculture all'Aperto.

Hewlett, P., ed. 2001. *Lea's Chemistry of Cement and Concrete*. 4th ed. Oxford: Butterworth-Heinemann.

Hoke, E. 1976. Microprobe investigations of incrusted as well as cleaned marble specimens. In *Second International Congress on Deterioration and Conservation of Stone*, ed. T. Skoulikidis, 119–26. Athens: Ministry of Culture and Science of Greece.

Horie, C. V. 1987. *Materials for Conservation*. London: Butterworths.

Iler, R. K. 1979. *The Chemistry of Silica*. New York: Wiley.

Jerome, P. S., N. R. Weiss, A. S. Gilbert, and J. A. Scott. 1998. Ethyl silicate treatment as a treatment for marble: Conservation of St. John's Hall, Fordham University. *APT Bulletin* 24 (1):19–26.

Keefer, K. D. 1984. The effect of hydrolysis conditions on the structure and growth of silicate polymers. In *Better Ceramics through Chemistry*, ed. C. J. Brinker, D. E. Clark, and D. R. Ulrich, 15–24. New York: North-Holland.

Khaskin, I. G. 1952. Several applications of deuterium and heavy oxygen in the chemistry of flint. *Dokl. Akad. Nauk SSSR* 85:129.

Kimmel, J., and G. Wheeler. n.d. The apse from San Martin de Fuentidueña: Condition history, survey of conditions, analyses. Unpublished manuscript [1999].

King, G. 1930. Silicon esters, and their application to the paint industry. *Journal of the Oil and Colour Chemists' Association* 13:28–55.

King, G. 1931. Silicon ester binder. *Paint Manufacture* 16–20.

King, G., and R. Threlfall. 1928. Improvements in materials for use in the art of painting or varnishing. U.K. Patent 290,717. 16 May.

King, G., and R. Threlfall. 1931. Material for forming coatings, for use as impregnating agents or for like purposes. U.S. Patent Application 1,809,755. 9 June.

King, G., and R. Threlfall. 1927. Verfahren zur Herstellung von Uberzugs-Impragnierungs-oder plastischen Massen. German Patent Application 553,514. 9 December.

King, G., and A. R. Warnes. 1938. Improvements relating to the treatment of porous surfaces such as that of stone, and to compositions therefor. U.K. Patent Application 699,238. 5 March.

King, G., and A. R. Warnes. 1939. Improvements relating to the treatment of porous surfaces such as that of stone, and to compositions therefor. U.K. Patent 513,366. 11 October.

Klein, L. C. 1985. Sol-gel processing of silicates. *Annual Review of Material Science* 15:227–48.

Klemm, D. D., R. Snethlage, and B. Graf 1977. Bericht über die Steinrestaurierung am Bayer-Tor in Landsberg am Lech. *Maltechnik-Restauro* 83:242–52.

Koblischek, P. 1995. The consolidation of natural stone with stone strengthener on the basis of poly-silicic-acid-ethyl-ester. In *Conservation et restauration des biens culturels*, ed. R. Pancella, 261–65. Lausanne: Ecole Polytechnique Federale de Lausanne.

Koestler, R. J. 2000. Polymers and resins as food for microbes. In *Of Microbes and Art: The Role of Microbial Communities in the Degradation and Protection of Cultural Property*, 153–66. New York: Kluwer.

Koestler, R. J., A. E. Charola, and G. Wheeler. 1985. Scanning electron microscopy in –conservation: The Abydos reliefs. In *Application of Science in Examination of Works of Art*, ed. P. A. England and L. van Zelst, 225–29. Boston: Museum of Fine Arts Research Laboratory.

Koestler, R. J., and E. D. Santoro. 1988. Assessment of the susceptibility to biodeterioration of selected polymers and resins. Technical Report for the Getty Conservation Institute, Marina del Rey, Calif.

Koestler, R. J., E. D. Santoro, F. Preusser, and A. Rodarte. 1986. A note on the reaction of methyltrimethoxysilane to mixed cultures and microorganisms. In *Biodeterioration Research 1*, ed. C. E. O'Rear and G. C. Lllewellyn, 317–21. New York: Plenum.

Kozlowski, R., M. Tokarz, and M. Persson. 1992. "Gypstop"—a novel protective treatment. In *Seventh International Congress on Deterioration and Conservation of Stone*, ed. J. Delgado Rodrigues, F. Henriques, and F. Jeremias, 1187–96. Lisbon: LNEC.

Kumar, R. 1995. Fourier transform infrared spectroscopic study of silane/stone interface. In *Materials Issues in Art and Archaeology IV*, ed. P. B. Vandiver, J. Druzik, J. M. L. Galvan, I. C. Freestone, and G. Wheeler, 341–47. Pittsburgh: Materials Research Society.

Kwiatkowski, D., and M. Klingspor. 1995. Consolidation of Gotland stone in monuments. In *Methods of Evaluating Products for the Conservation of Porous Building Materials in Monuments*, ed. M. L. Tabasso, 170–87. Rome: ICCROM.

Ladenberg, A. 1874. Ueber neue Siliciumverbindung. *Annalen der Chemie* 173:143–66.

Larson, J. 1980. The conservation of stone sculptures in historic buildings. In *Conservation within Historic Buildings*, ed. N. Brommelle, G. Thomson, and P. Smith, 132–38. London: IIC.

Larson, J. 1983. A museum approach to the techniques of stone conservation. In *Fourth International Congress on Deterioration and Preservation of Stone Objects*, ed. K. L. Gauri and J. A. Gwinn, 219–38. Louisville: University of Louisville.

Laurie, A. P. 1923. Improvements relating to the preservation of stone. U.K. Patent Application 203,042. 6 September.

Laurie, A. P. 1924. Preservation of stone. U.S. Patent Application 732,574. 16 August.

Laurie, A. P. 1925. Preservation of stone. U.S. Patent 1,561,988. 17 November.

Laurie, A. P. 1926a. Preservation of stone. U.S. Patent 1,585,103. 18 May.

Laurie, A. P. 1926b. Preservation of Stone. U.S. Patent 1,607,762. 23 November.

Leisen, H., J. Poncar, and S. Warrack. 2000. *German Apsara Conservation Project*. Köln: German Apsara Conservation Project.

Lerner, R. W., and A. R. Anderson. 1967. Silicon-containing water repellent compositions. U.S. Patent 3,310,417. 21 May.

Lewin, S. Z. 1996a. *The Preservation of Natural Stone, 1839–1965: An Annotated Bibliography. Art and Archaeology Technical Abstracts* 6 (1) suppl.

Lewin, S. Z . 1966b. The preservation of stone. U.S. Patent Application 529,213.

Lewin, S. Z. 1972. Recent experience with chemical techniques of stone preservation. In *The Treatment of Stone*, ed. R. Rossi-Manaresi and G. Torraca, 139–44. Bologna: Centro per la Conservazione delle Sculture all'Aperto.

Liberti, S. 1955. Consolidamento dei materiali da costruzione di monumenti antichi. *Bolletino dell'Istituto Centrale del Restauro* 21–22:43–70.

Lindborg, U., R. C. Dunakin, and D. Rowcliffe. 2000. Thermal stress and weathering of Carrara, Pentelic, and Ekeberg marble. In *Ninth International Congress on Deterioration and Conservation of Stone*, vol. 1, ed. V. Fassina, 109–17. Amsterdam: Elsevier.

Luckat, S. 1972. Investigations concerning the protection against air pollutants of objects of natural stone. *Staub-Reinhaltung der Luft* 32 (5):30–33.

Lukaszewicz, J. W. 1994. The application of silicone products in the conservation of volcanic tuffs. In *Lavas and Volcanic Tuffs*, ed. A. E. Charola, R. J. Koestler, and G. Lombardi, 191–202. Rome: ICCROM.

Lukaszewicz, J. W. 1996a. The influence of stone pre-consolidation with ethyl silicate on deep consolidation. In *Eighth International Congress on Deterioration and Conservation of Stone*, ed. J. Riederer, 1209–14. Berlin: Möller Druck und Verlag.

Lukaszewicz, J. W. 1996b. The influence of pre-consolidation with ethyl silicate on soluble salts removal. In *Eighth International Congress on Deterioration and Conservation of Stone*, ed. J. Riederer, 1203–8. Berlin: Möller Druck und Verlag.

Lukaszewicz, J. W., D. Kwiatkowski, and M. Klingspor. 1995. Consolidation of Gotland stone in monuments. In *Methods of Evaluating Products for the Conservation of Porous Building Materials in Monuments,* ed. M. L. Tabasso, 170–87. Rome: ICCROM.

Mangio, R., and A.-M. Lind. 1997. Silica SOLS (gypstop) for simultaneous desalination and consolidation of Egyptian limestone at the Karnak Temple in Luxor, Egypt. In *Fourth International Symposium for the Conservation of Monuments in the Mediterranean Basin,* ed. A. Moropoulou, F. Zezza, E. Kollias, and E. Papachristodoulou, 299–312. Athens: Technical Chamber of Greece.

Mangum, B. 1986. On the choice of preconsolidant in the treatment of an Egyptian polychrome triad. In *Case Studies in the Conservation of Stone and Wall Paintings,* ed. N. S. Brommelle and P. Smith, 148–50. London: IIC.

Martin, B., D. Mason, J. M. Teutonico, and S. Chapman. 2002. Stone consolidants: Brethane report on an 18-year review of Brethane-treated sites. In *Stone: Stone Building Materials, Construction and Associated Component Systems. Their Decay and Treatment,* ed. J. Fidler, 3–18. English Heritage Research Transactions 2. London: James & James.

Martin, W. 1996. Stone consolidants: A review. In *A Future for the Past,* ed. J. M. Teutonico, 30–49. London: James & James.

Matero, F., and A. Oliver. 1997. A comparative study of alkoxysilanes and acrylics in sequence and in mixture. *Journal of Architectural Conservation* 3 (2):22–42.

Miller, E. 1992. Current practice at the British Museum for the consolidation of decayed porous stones. *Conservator* 16:78–83.

Moncrieff, A. 1976. The treatment of deteriorating stone with silicone resins: Interim report. *Studies in Conservation* 21 (4):179–91.

Müller, R. 1965. One hundred years of organosilicon chemistry. Trans. E. G. Rochow. *Journal of Chemical Education* 42 (1):41–47.

Munnikendam, R. A. 1967. Preliminary notes on the consolidation of porous building materials by impregnation with monomers. *Studies in Conservation* 12 (4):158–62.

Munnikendam, R. A. 1971. Acrylic monomer systems for stone impregnation. In *La Conservazione delle sculture all'aperto,* ed. R. Rossi-Manaresi and D. Mietto, 221–28. Bologna: Edizione Alfa.

Munnikendam, R. A. 1972. The combination of low viscosity epoxy resins and silicon-esters for the consolidation of stone. In *The Treatment of Stone,* ed. R. Rossi-Manaresi and G. Torraca, 197–200. Bologna: Centro per la Conservazione delle Sculture all'Aperto.

Nagy, K. L., R. Cygan, C. S. Scotto, C. J. Brinker, and C. S. Ashley. 1997. Use of coupled passivants and consolidants on calcite mineral surfaces. In *Materials Issues in Art and Archaeology V: Symposium Held December 3–5, 1996, Boston, Massachusetts, USA,* ed. J. F. Merkel and J. Stewart, 301–6. Pittsburgh: Materials Research Society.

Nishiura, T. 1986. Study on the color change induced by the impregnation of silane. I. Analysis of the color change by color meter. In *Study on the Conservation Treatment of Stone VII. Scientific Papers on Japanese Antiques and Art Crafts,* 31, 41–50.

Nishiura, T. 1987a. Laboratory evaluation of the mixture of silane and organic resin as consolidant of granularly decayed stone. In *Preprints of the ICOM Eighth Triennial Meeting in Sydney, Australia,* ed. K. Grimstad, 805–7. Los Angeles: Getty Conservation Institute.

Nishiura, T. 1987b. Laboratory test on the color change of stone by impregnation with silane. In *Preprints of the ICOM Eighth Triennial Meeting in Sydney, Australia,* ed. K. Grimstad, 509–12. Los Angles: Getty Conservation Institute.

Nishiura, T. 1995. Experimental evaluation of stone consolidants used in Japan. In *Methods of Evaluating Products for the Conservation of Porous Building Materials in Monuments,* ed. M. L. Tabasso, 189–202. Rome: ICCROM.

Nishiura, T., M. Fukuda, and S. Miura. 1984. Treatment of stone with synthetic resins for its protection against damage by freeze-thaw cycles. In *Adhesives and Consolidants,* ed. N. S. Brommelle, E. M. Pye, P. Smith, and G. Thomson, 156–59. London: IIC.

Noll, W. 1968. *Chemistry and Technology of Silicones.* New York: Academic Press. Originally published in German by Verlag Chemie.

Nonfarmale, O. 1976. A method of consolidation and restoration for decayed sandstones. In *The Conservation of Stone I,* ed. R. Rossi-Manaresi, 401–10. Bologna: Centro per la Conservazione delle Sculture all'Aperto.

North American Graduate Programs in the Conservation of Cultural Property. 2000. Buffalo, N.Y.: ANAGPIC (Petit Printing Corporation).

Nunberg, S., A. Heywood, and G. Wheeler. 1996. Relative humidity control as an alternative approach to preserving an Egyptian limestone relief. In *Le dessalement des matériaux poreux,* 127–35. Seventh Journées d'Etudes de La SFIIC. Champs-sur-Marne: SFIIC.

Oliver, A. B. 2002. The variable performance of ethyl silicate: Consolidated stone at three National Parks. *APT Bulletin* 33 (2–3):39–44.

Partington, J. R. 1964. *A History of Chemistry.* London: Macmillan.

Peeler, R. L., and S. A. Kovacich. 1959. Alkyl silicate aviation hydraulic fluids. *Industrial Engineering Chemistry* 51:740–52.

Perez, J. L., R. Villegas, J. F. Vale, M. A. Bello, and M. Alcalde. 1995. Effects of consolidant and water repellent treatments on the porosity and pore size distribution of limestones. In *Methods of Evaluating Products for the Conservation of Porous Building Materials in Monuments,* ed. M. L. Tabasso, 203–11. Rome: ICCROM.

Pettijohn, E. J. 1975. *Sedimentary Rocks.* 3d ed. New York: Harper and Row.

Phillips, M. W. 1982. Acrylic precipitation consolidants. In *Science and Technology in the Service of Conservation,* ed. N. S. Brommelle and G. Thomson, 56–60. London: IIC.

Plehwe-Leisen, E. von, E. Wendler, H. D. Castello Branco, and A. F. Dos Santos. 1996. Climatic influences on long-term efficiency of conservation agents for stone: A German-Brazilian outdoor exposure program. In *Eighth International Congress on Deterioration and Conservation of Stone,* ed. J. Riederer, 1325–32. Berlin: Möller Druck und Verlag.

Plenderleith, H. J. 1956. *The Conservation of Antiquities and Works of Art.* London: Oxford University Press.

Plenderleith, H. J., and A. E. A. Werner. 1971. *The Conservation of Antiquities and Works of Art.* 2d ed. London: Oxford University Press.

Plueddemann, E. 1991. *Silane Coupling Agents.* 2d ed. New York: Plenum.

Pouxviel, J. C., J. P. Boilet, J. C. Beloeil, and J. Y. Lallemand. 1987. NMR study of sol/gel polymerization. *Journal of Non-Crystalline Solids* 89:345.

Press, F., and R. Siever. 1978. *Earth.* 2d ed. San Francisco: W. H. Freeman.

Price, C. A. 1975. The decay and preservation of natural building stone. Building Establishment Current Paper, Reprint II (CP89/75):350–53.

Price, C. A. 1996. *Stone Conservation: An Overview of Current Research.* Marina del Rey, Calif.: Getty Conservation Institute.

Price, C., and P. Brimblecombe. 1994. Preventing salt damage in porous materials. In *Preventive Conservation Practice, Theory and Research,* ed. A. Roy and P. Smith, 90–93. London: IIC.

Rager, G., M. Payre, and L. Lefèvre. 1996. Mise au point d'une méthode de dessalement pour des sculptures du XIVᵉ siècle en pierre polychromée. In *Le dessalement des matériaux poreux,* 241–56. 7th Journées d'Etudes de la SFIIC. Champs-sur-Marne: SFIIC.

Rao, S. M., C. J. Brinker, and T. J. Ross. 1996. Environmental microscopy in stone conservation. *Scanning* 18:508–14.

Riddick, J., W. Bunger, and T. Sakano. 1986. *Organic Solvents: Physical Properties and Methods of Purification.* 4th ed. New York: Wiley.

Riederer, J. 1971. Stone preservation in Germany. In *New York Conference on the Conservation of Stone and Wooden Objects,* 2d ed., vol. 1, ed. G. Thomson, 125–34. London: IIC.

Riederer, J. 1972a. Steinkonservierung im lichte neuerer erkenntnisse. *Eurafem-Information* 3:1–12.

Riederer, J. 1972b. Stone conservation with silicate esters. Paper presented at the meeting of the ICOM Committee for Conservation, Madrid, 2–8 October. Mimeograph.

Riederer, J. 1973. Steinkonservierung am Aphaia-Tempel auf Aegina. *Maltechnik-Restauro* 79:193–99.

Riederer, J. 1974. Die Erhaltung Ägyptischer Baudenkmaler. *Maltechnik-Restauro* 80:43–52.

Roby, T. 1996. In-situ assessment of surface consolidation and protection treatments of marble monuments in Rome in the 1980s, with particular reference to two treatments with B72. In *Eighth International Congress on Deterioration and Conservation of Stone,* ed. J. Riederer, 1015–28. Berlin: Möller Druck und Verlag.

Rochow, E. G. 1995. Why silicon? In *Progress in Organosilicon Chemistry,* ed. B. Marciniec and J. Chojnowski, 3–15. Basel: Gordon and Breach.

Rodrigues, J. Delgado, and D. Costa, eds. 1996. *Conservation of Granitic Rocks.* Lisbon: LNEC.

Rodrigues, J. Delgado, D. Costa, M. Sá da Costa, and I. Eusébio. 1996. Behavior of consolidated granites under aging tests. In *Conservation of Granitic Rocks,* ed. J. Delgado Rodrigues and D. Costa, 79–85. Lisbon: LNEC.

Rodriguez-Navarro, C., E. Hansen, E. Sebastian, and W. S. Ginell. 1997. The role of clays in the decay of ancient Egyptian limestone sculpture. *Journal of the American Institute for Conservation* 36 (2):151–63.

Rohatsch, A., J. Nimmrichter, and I. Chalupar. 2000. Physical properties of fine-grained marble before and after conservation. In *Ninth International Congress on Deterioration and Conservation of Stone,* ed. V. Fassina, 453–58. Amsterdam: Elsevier.

Rolland, O., P. Floc'h, G. Martinet, and V. Verges-Belmin. 2000. Silica bound mortars for the repairing of outdoor granite sculptures. In *Ninth International Congress on Deterioration and Conservation of Stone,* vol. 2, ed. V. Fassina, 307–15. Amsterdam: Elsevier.

de Ros, D., and F. Barton. 1926. Improved methods of hardening and preserving natural and artificial stones. U.K. Patent 260,031. 18 October.

Rossi-Manaresi, R. 1976. Treatments for sandstone consolidation. In *The Conservation of Stone I,* ed. R. Rossi-Manaresi, 547–72. Bologna: Centro per la Conservazione delle Sculture all'Aperto.

Rossi-Manaresi, R., ed. 1981. *The Conservation of Stone II.* 2 vols. Bologna: Centro per la Conservazione delle Sculture all'Aperto.

Rossi-Manaresi, R. 1986. *Conservation Works in Bologna and Ferrara.* London: IIC.

Rossi-Manaresi, R., and G. Chiari. 1980. Effectiveness of conservation treatments of a volcanic tuff very similar to adobe. In *Third International Symposium on Mud-Brick (Adobe) Preservation,* 29–38. Ankara: ICOM-ICOMOS.

Rossi-Manaresi, R., A. Rattazzi, and L. Toniolo. 1995. Long-term effectiveness of treatments of sandstone. In *Methods of Evaluating Products for the Conservation of Porous Building Materials in Monuments,* ed. M. L. Tabasso, 225–44. Rome: ICCROM.

Ruedrich, T., T. Weiss, and S. Siegesmund. 2002. Thermal behavior of weathered and consolidated marbles. In *Natural Stone, Weathering Phenomena, Conservation Strategies and Case Studies,* ed. S. Siegesmund, T. Weiss, and A. Vollbrecht, 255–72. Geological Society Special Publication No. 205. London: Geological Society.

Ruggieri, G., E. Cajano, G. Delfini, P. Mora, L. Mora, and G. Torraca. 1991. Il Restauro conservativo della facciata di S. Andrea della valle in Roma. In *Le Pietre nell'architettura struttura e superfici: Atti del convegno di studi Bressanone 25–28 giugo 1991,* ed. G. Biscontin and D. Mietto, 535–44. Scienza e beni culturali 7. Padua: Libreria Progetto Editore.

Sage, J. D. 1988. Thermal microfracturing of marble. In *The Engineering Geology of Ancient Works, Monuments and Historic Sites,* vol. 2, ed. P. Marinos and G. Koukis, 1013–18. 4 vols. Rotterdam: A. A. Balkema.

Saiz-Jimenez, C. 1991. Characterization of organic compounds in weathered stone. In *Science, Technology and European Cultural Heritage,* ed. N. Baer, C. Sabbioni, and A. I. Sors, 523–26. Oxford: Butterworths.

Sakka, S. 1984. Formation of glass and amorphous oxide fibers from solution. In *Better Ceramics through Chemistry,* ed. C. J. Brinker, D. E. Clark, and D. R. Ulrich, 91–99. New York: North-Holland.

Saleh, S. A., F. M. Helmi, M. M. Kamal, and A. E. El-Banna. 1992a. Artificial weathering of treated limestone: Sphinx, Giza, Egypt. In *Seventh International Congress on Deterioration and Conservation of Stone,* ed. J. Delgado Rodrigues, F. Henriques, and F. Telmo Jeremias, 781–89. LNEC.

Saleh, S. A., F. M. Helmi, M. M. Kamal, and A. E. El-Banna. 1992b. Study and consolidation of sandstone: Temple of Karnak, Luxor, Egypt. *Studies in Conservation* 37 (2):93–104.

Sasse, H. R., D. Honsinger, and B. Schwamborn. 1993. PINS: A new technology in porous stone conservation. In *The Conservation of Stone and Other Materials,* vol. 2, ed. M. -J. Thiel, 705–16. New York: E. & F. N. Spon.

Sattler, L., and H. Schuh. 1995. Zur zeitlichen Entwicklung von Steinfestigungen auf Kieselsäureesterbasis. *Bautenschutz und Bausanierung* 1:77–81.

Sattler, L., and R. Snethlage. 1988. Durability of stone consolidation treatments with silicic acid ester. In *The Engineering Geology of Ancient Works, Monuments and Historical Sites,* vol. 2, ed. P. G. Marinos and G. C. Koukis, 953–56. Rotterdam: A. A. Balkema.

Schaffer, R. J. [1932] 1972. *The Weathering of Natural Building Stones.* Watford: BRE.

Scherer, G., and G. Wheeler. 1997. Stress development drying of *Conservare* OH. In *Fourth International Symposium on the Conservation of Monuments in the Mediterranean Basin,* ed. A. Moropoulou, F. Zezza, E. Kollias, and I. Papachristodoulou, 355–62. Athens: Technical Chamber of Greece.

Schmidt, H., and D. Fuchs. 1991. Protective coatings for medieval stained glass. In *Materials Issues in Art and Archaeology II,* ed. P. B.Vandiver, J. Druzik, and G. Wheeler, 227–38. Pittsburgh: Materials Research Society.

Schmidt, H., H. Scholze, and A. Kaiser. 1984. Principles of hydrolysis and condensation of alkoxysilanes. *Journal of Non-Crystalline Solids* 63:1–11.

Shaw, C., and J. E. Hackford. 1945. Development in the application of silicic esters. *Industrial Chemist* 21 (March):130–35.

Shore, B. C. G. 1957. *Stones of Britain.* London: Leonard Hill Books.

Silicones for Stone Conservation. 1999. Wacker-Chemie Silicones Division product literature, no. 5496e, 1.99.

Sleater, G. A. 1977. Stone preservatives: Methods of laboratory testing and preliminary performance criteria. *NBS Technical Note 941.* Washington, D.C.

Smith, A. L., ed. 1983. *Analysis of Silicones.* Malabar, Fla.: Krieger.

Smith, D. 1957. Cleaning inscriptions and sculptures in sandstone. *Museums Journal 57*, 1957:215–19.

Smith, D. M., G. Scherer, and J. Anderson. 1995. Shrinkage during drying of silica gel. *Journal of Non-Crystalline Solids* 188:191–206.

Snethlage, R. 1996. Der Grosse Buddha von Dafosi. *Arbeitshefte des Bayerischen Landesamtes für Denkmakpflege* 82:220–39.

Snethlage, R. 1983. Die Wasserdampf-Sorptionseigenshcaften von Sandsteinen und ihre Bedeutung fur die Konservierung. In *Werkstoffwissenshcanften und Bausanierung,* ed. F. H. Wittmann, 297–303. Stuttgart: Elvira Moeller.

Snethlage, R., and D. D. Klemm. 1978. Scanning electron microscopic investigation on impregnated sandstones. *Deterioration and Protection of Stone Monuments,* vol. 2, no. 5.7. Paris: UNESCO.

Snethlage, R., and E. Wendler. 1991. Surfactants and adherent silicon resins—new protective agents for natural stone. In *Materials Issues in Art and Archaeology II,* ed. P. Vandiver, J. Druzik, and G. Wheeler, 193–200. Pittsburgh: Materials Research Society.

Sneyers, R. V. 1963. 2e rapport sur l'étude des matériaux pierreux. Report presented at ICOM meeting, Leningrad and Moscow, 16–23 September. Mimeograph.

Sneyers, R. V., and P. J. de Henau. 1968. The conservation of stone. In *The Conservation of Cultural Property,* 230–31. Paris: UNESCO Press. [Reprint 1975.]

Sramek, J., and V. Eckert. 1986. The Kohl Fountain in Prague—conservation of a stone object in prolonged contact with water. In *Case Studies in the Conservation of Stone and Wall Paintings,* ed. N. S. Brommelle and P. Smith, 109–11. London: IIC.

Stadlbauer, E., S. Lotzmann, B. Meng, H. Rösch, and E. Wendler. 1996. On the effective-ness of stone conservation after 20 years of exposure—case study of Clemenswerth Castle/NW-Germany. In *Eighth International Congress on Deterioration and Conservation of Stone,* ed. J. Riederer, 1285–96. Berlin: Möller Druck und Verlag.

Stepien, P. 1988. Case studies in the use of organo-silanes as consolidants in conjunction with traditional lime technology. In *Sixth International Congress on Deterioration and Conservation of Stone,* ed. J. Ciabach, 647–52. Torun: Nicholas Copernicus University.

Stone, J. B., and A. J. Teplitz. 1942. Earth consolidation. U.S. Patent 2,281,810. 5 May.

Swern, D., ed. 1979. *Bailey's Industrial Oil and Fat Products*, vol. 1. 4th ed. New York: Wiley.

Tabasso, M. L., and U. Santamaria. 1985. Consolidant and protective effects of different products on Lecce limestone. In *Fifth International Congress on Deterioration and Conservation of Stone,* ed. G. Felix and V. Furlan, 697–707. Lausanne: Presses Polytechniques Romandes.

Tabasso, M. L., A. M. Mecchi, and U. Santamaria. 1994. Interaction between volcanic tuff and products used for consolidation and waterproofing treatments. In *Lavas and Volcanic Tuffs,* ed. A. E. Charola, R. J. Koestler, and G. Lombardi, 173–90. Rome: ICCROM.

Thickett, N. L., and S. M. Bradley. 2000. Assessment of the performance of silane treatments applied to Egyptian limestone sculptures displayed in a museum environment. In *Ninth International Congress on Deterioration and Conservation of Stone*, ed. V. Fassina, 503–11. Amsterdam: Elsevier.

Thomson-Houston Company. 1947, 1948. Improvements in and relating to methods of preparing alkyl-orthosilicates. U.K. Patent Application 16234/47, 19 June. U.K. Patent 629,138. 13 September.

Tomkeieff, S.I. 1983. *Dictionary of Petrology*. New York: Wiley.

Tudor, P. B. 1989. The restoration of a marble statue in New Orleans, Louisiana: A case study. Master's thesis, Columbia University.

Useche, L. A. 1994. Studies for the consolidation of the façade of the church of Santo Domingo, Popayan, Colombia. In *Lavas and Volcanic Tuffs*, ed. A. E. Charola, R. J. Koestler, and G. Lombardi, 165–71. Rome: ICCROM.

Vega, A., and G. Scherer. 1989. Study of structural evolution of silica gel using ^{1}H and ^{29}Si NMR. *Journal of Non-Crystalline Solids* 111:153–66.

Verges-Belmin, V., D. Garnier, A. Bouineau, and R. Coignard. 1991. Impregnation of badly decayed Carrara marble by consolidating agents: Comparison of seven treatments. In *Second International Symposium on the Conservation of Monuments in the Mediterranean Basin*, ed. D. Decrouez, J. Chamay, and F. Zezza, 421–37. Geneva: Ville de Genève Muséum d'Historique naturelle & Musée d'art et d'histoire.

Voronkov, M. G., V. P. Mileshkevich, and Y. A. Yuzhelevski. 1978. *The Siloxane Bond*. New York: Consultants Bureau.

Wagner, H. 1956. *Taschenbuch des Chemischen Bautenschutzes*. 4th ed. Stuttgart: Wissenschaftliche Verlagsgesellschaft.

Weber, H. 1976. Stone renovation and consolidation using silicones and silicic esters. In *The Conservation of Stone I*, ed. R. Rossi-Manaresi, 375–85. Bologna: Centro per la Conservazione delle Sculture all'Aperto.

Weber, H. 1977. Uraschen und Behandlung der Steinverwitterung. *Maltechnik-Restauro* 83:73–89.

Weber, H. 1985. The conservation of the Alte Pinakothek in Munich. In *Fifth International Congress on Deterioration and Conservation of Stone*, ed. G. Felix and V. Furlan, 1063–72. Lausanne: Presses Polytechniques Romandes.

Weeks, C. 1998. The Portail de la Mere Dieu of Amiens Cathedral: Its polychromy and conservation. *Studies in Conservation* 43 (2):101–8.

Weiss, N. R., I. Slavid, and G. Wheeler. 2000. Development and assessment of a conversion treatment for calcareous stone. In *Ninth International Congress on Deterioration and Conservation of Stone*, vol. 2, ed. V. Fassina, 533–40. Amsterdam: Elsevier.

van der Weij, F. W. 1980. The action of tin compounds in condensation-type RTV silicone rubbers. *Makromolecular Chemie* 181:2541–48.

Wendler, E. 1996. New materials and approaches for the conservation of stone. In *Saving Our Architectural Heritage*, ed. N. Baer and R. Snethlage, 182–96. New York: Wiley.

Wendler, E., A. E. Charola, and B. Fitzner. 1996. Easter Island tuff: Laboratory studies for its consolidation. In *Eighth International Congress on Deterioration and Conservation of Stone*, ed. J. Riederer, 1159–70. Berlin: Möller Druck und Verlag.

Wendler, E., D. Klemm, and R. Snethlage. 1991. Consolidation and hydrophic treatments of natural stone. In *The Durability of Building Materials and Components*, ed. J. M. Baker, A. J. Majumdar, and H. Davies, 203–12. Brighton: Chapman & Hall.

Wendler, E., and L. Sattler. 1996. Bohrwiderstandsmessung als zerstörungsarmes Prüfverfahren zur Bestimmung des Festigkeitsprofils in Gesteinen und Keramik. In *Fourth International Colloquium "Werkstoffwissenschaften und Bausanierung."* Esslingen: TAE.

Wendler, E., L. Sattler, S. Simon, H. Ettl, and F. Heckmann. 1999. Untersuchungen zur Wirksamkeit von Konsolidierungmitteln an Tuffeaustein von Kathedrale St. Gatien in Tours. In *Monuments historiques et environments*, 244–61. Colloquium on Recherches Franco-Allemandes sur la pierre el le vitrail 1988–1996. Paris: Exé Productions.

Wendler, E., L. Sattler, P. Zimmermann, D. D. Klemm, and R. Snethlage. 1992. Protective treatment of natural stone: Requirements and limitations with respect to the state of damage. In *Seventh International Congress on Deterioration and Conservation of Stone*, ed. J. Delgado Rodrigues, F. Henriques, and F. Jeremias, 1103–12. Lisbon: LNEC.

Wheeler, G., J. Dinsmore, L. J. Ransick, A. E. Charola, and R. J. Koestler. 1984. Treatment of the Abydos reliefs: Consolidation and cleaning. *Studies in Conservation* 29 (1):42–48.

Wheeler, G., S. A. Fleming, and S. Ebersole. 1991. Evaluation of some current treatments for marble. In *Second International Symposium on the Conservation of Monuments in the Mediterranean Basin*, ed. D. Decrouez, J. Chamay, and F. Zezza, 439–43. Geneva: Ville de Genève, Muséum d'Histoire naturelle & Musée d'art et d'histoire.

Wheeler, G., S. A. Fleming, and S. Ebersole. 1992. Comparative strengthening effect of several consolidants on Wallace sandstone and Indiana limestone. In *Seventh International Congress on Deterioration and Conservation of Stone*, ed. J. Delgado Rodrigues, F. Henriques, and F. Jeremias, 1033–41. Lisbon: LNEC.

Wheeler, G., J. Mendez-Vivar, E. S. Goins, S. A. Fleming, and C. J. Brinker. 2000. Evaluation of alkoxysilane coupling agents in the consolidation of limestone. In *Ninth International Congress on Deterioration and Conservation of Stone*, vol. 2, ed. V. Fassina, 541–45. Amsterdam: Elsevier.

Wheeler, G., G. L. Shearer, S. Fleming, L. W. Kelts, A. Vega, and R. J. Koestler. 1991. Toward a better understanding of ACRYLOID B72 acrylic resin/methyltrimethoxysilane stone consolidants. In *Materials Issues in Art and Archaeology II*, ed. P. B. Vandiver, J. R. Druzik, and G. Wheeler, 209–32. Pittsburgh: Materials Research Society.

Wheeler, G., and R. Newman. 1994. Analysis and treatment of a stone urn from the Imperial Hotel, Tokyo. In *Lavas and Volcanic Tuffs*, ed. A. E. Charola, R. J. Koestler, and G. Lombardi, 157–61. Rome: ICCROM.

Wheeler, G., A. Schein, G. Shearer, S. H. Su, and C. S. Blackwell. 1992. Preserving our heritage in stone. *Analytical Chemistry* 64 (5):347–56.

Wihr, R. 1976. Deep impregnation for effective stone protection. In *The Conservation of Stone I*, ed. R. Rossi-Manaresi, 317–18. Bologna: Centro per la Conservazione delle Sculture all'Aperto.

Wihr, R. 1978. The use of aethyl-silicate and acrylic-monomers in stone preservation. In *Preprints for the International Symposium: Deterioration and Protection of Stone Monuments, 5–9 June 1978*, vol. 3, no. 7.12. Paris: RILEM and UNESCO.

Willensky, E., and N. White. 1988. *The AIA Guide to New York City.* 3d ed. New York: Harcourt Brace Jovanovich.

Winkler, E. M. 1994. *Stone in Architecture.* 3d ed. New York: Springer-Verlag.

Wittman, F. H., and R. Prim. 1983. Mesures de l'effet consolidant d'un produit de traitement. *Matériaux et Construction* 16(94):235–42.

Worch, E. 1973. Steinkonservierung. *Th. Goldschmidt* 3 (24):24–26.

Zanardi, B., L. Calzetti, A. Casoli, A. Mangia, G. Rizzi, and S. Volta. 1992. Observations on a physical treatment of stone surfaces. In *Seventh International Congress on Deterioration and Conservation of Stone*, ed. J. Delgado Rodrigues, F. Henriques, and F. Jeremias, 1243–51. Lisbon: LNEC.

Annotated Bibliography

Elizabeth Stevenson Goins
With contributions from George Wheeler, Carol A. Grissom, and
Norman R. Weiss

The purpose of this bibliography is to update Carol A. Grissom and Norman R. Weiss's *Alkoxysilanes in the Conservation of Art and Architecture: 1861-1981*, the 1981 supplement to *Art and Archaeology Technical Abstracts*. Some entries have been added and some dropped for the period prior to 1981, and most of the pre-1981 abstracts have been shortened.

The bibliography attempts to be complete (i.e., every article or book on alkoxysilanes as it relates to stone consolidants should appear below). This is the goal we strive for, knowing also that we will fail. However, after the publication of this book, this bibliography will continue to have a life as the first new supplement to be added to AATA Online (www.aata.getty.edu), which is updated quarterly.

1824
Berzelius, J. J. Untersuchungen über die Flusspathsäure und deren merkwürdigsten Verbindungen. *Annalen der Physik und Chemie* 1:169–230. Synthesis of silicon tetrachloride.

1846
Ebelmen, J. J. von. Untersuchungen über die Verbindungen der Borsäure mit Aether. *Annalen der Chemie und Pharmacie* 57:319–53. Synthesis of tetraethoxysilane.

1861
Anon. Stone-preserving processes: Royal Institute of British Architects. *The Builder* 19:103–5. "Silicic ether" is suggested as a potential treatment for the Houses of Parliament.

1874
Ladenberg, A. Ueber neue Siliciumverbindung. *Annalen der Chemie* 173:143–66. Synthesis of methyltriethoxysilane.

1923
Laurie, A. P. Improvements relating to the preservation of stone. U.K. Patent 203,042, issued 6 September 1923. Claims use of "silicic ether" for the preservation of stone.

1925
Laurie, A. P. Preservation of stone. U.S. Patent 1,561,988, issued 17 November 1925. Description of the invention process. This is the only U.S. version to state that it is essential to dilute the silicic ester with a solvent to allow the solution to penetrate the stone.

1926
de Ros, D., and F. Barton. Improved methods of hardening and preserving natural and artificial stones. U.K. Patent Application 18379, 18 July 1925, U.K. Patent 260,031,

issued 18 October 1926. Silicon compounds ($SiCl_4$, SiF_4 . . .) are stated to strengthen stone.

Laurie, A. P. *The Painter's Methods and Materials*. Philadelphia: J. B. Lippincott, 1926. Reprint New York: Dover Publications, 1967. Suggests "silicon ester" for the preservation of wall paintings.

Laurie, A. P. Preservation of stone. U.S. Patent Application 732,574, 16 August 1924. U.S. Patent 1,585,103, issued 18 May 1926. Describes a "stone cement" based on "silicic ester" and volatile solvents. The solution may be applied to sandstone as is, and, after exposure to the atmosphere, the silicic ester hydrolyzes to form hydrated silica that cements stone particles together and gives partial waterproofing. Alkaline conditions before and during hydrolysis produce a soft, gelatinous precipitate that is useless as a cement. Acidic conditions, on the other hand, produce a hard, glassy substance. An acid catalyst is added to solutions applied to calcareous stones to compensate for this effect.

Laurie, A. P. Preservation of stone. U.S. Patent 1,607,762, issued 23 November 1926. A refined version of previous patents that focuses on the solution makeup and stresses that the "silicon ester" solution should be free of HCl.

1928

King, G., and R. Threlfall. Improvements in materials for forming coatings, for use as impregnating agents or for like purposes. U.K. Patent 290,717, issued 16 May 1928. Same as German Patent 553,514 (1932).

King, G., and R. Threlfall. Improvements in materials for use in the art of painting or varnishing. Great Britain Patent 290,717, issued 16 May 1928. Describes a mixture of an ester (or esters) of silicic acid, alcohol, water, and pigment as a paint medium. "Hydrolysis/coagulation rates" are found to be affected by water content, acidity, or alkalinity. The acid or base nature of the pigment, or the surface, is a factor to be considered.

1929

Oil and Colour Chemists' Association. *Oil and Colour Trades Journal* 76 (1927–32). Report on a paper presented by Mr. King, "Silicon esters and their application to the paint industry." Remarks on the need to use a mutual solvent to effect miscibility between ethyl silicate and water. The amount of water must be controlled and added to the solution—not by relying on water at the stone's surface. Reviews "silicon ester" paint medium (production and properties) that is made by a two-step, noncatalyzed process. Hardening of the film should be as slow as possible for best results (up to 4 months), and alkaline surfaces do "not influence unfavourably" the hardening process or film formation. However, the addition of alkaline materials (i.e., pigments) prior to application will cause the solution to gel prematurely. Also mentions use as stone preservative.

1930

Heaton, N. The possibilities of inorganic paint vehicles. *Journal of the Oil and Colour Chemists' Association* 13:330–40. Reports on the development of "silicon ester" paints.

King, G. Silicon esters, and their application to the paint industry. *Journal of the Oil and Colour Chemists' Association* 13:28–55. Ethyl silicates may be used for stone preservation by the precipitation of colloidal silica from ethyl silicate/alcohol solutions (usually containing 9% silica). This formulation is said to be suitable for old plaster, stone, and other porous surfaces. Notes that some pigments, like zinc oxide and basic pigments, cannot be added as the "paint will set to a solid in the can."

1931

King, G. Silicon ester binder. *Paint Manufacture*, pp. 16–20. "Silica and alcohol" solutions are made up that differ in the total silica content. Soluble substances (e.g., linseed oil) may be added to the solution that are distributed throughout the gel and may affect the mechanical properties.

King, G., and R. Threlfall. [1928.] Material for forming coatings, for use as impregnating agents or for like purposes. U.S. Patent Application 1,809,755, 9 June 1931. Identical to Great Britain Patent 290,717 (1928).

1932

Gardner, H. A., and G. G. Sward. Experiments to preserve fresco for exterior decoration. *American Paint and Varnish Manufacturer's Association Circular*, no. 421, pp. 285–89. Applied ethyl silicate solution to fresco panels. Ethyl silicate is added to alcohol and stirred until the solution is homogeneous, after which more ethyl silicate is added.

King, G., and R. Threlfall. Verfahren zur Herstellung von Überzugs-Imprägnierungs-oder plastischen Massen. German Patent 553,514, 27 June 1932. Concerns the use of "silicic acid ester" for manufacture of paints and coatings. Same as U.K. Patent 290,717 (1928).

Schaffer, R. J. *The Weathering of Natural Building Stones*. Building Research report no. 18. London: Dept. of Scientific and Industrial Research, 1932. Facsimile reprint Garston: BRE, 1972. Report on weathering tests conducted with a number of consolidants including "silicon esters."

1933

Graulich, W. Kieselsäure-Ester als Lackfarben-Grundkörper. *Nitrocellulose*, no. 4:61–62. Brief description of ethyl silicates and their use as stone preservatives.

Graulich, W. Zur Frage des Bautenschutzes: Neuzeitliche Steinkonservierung mit Kieselsäureester-Farben. *Tonindustrie-Zeitung und Keramische Rundschau* 57:677. Review of ethyl silicate paints.

1935

Ellis, C. Inorganic resins. In *The Chemistry of Synthetic Resins*, vol. 2:1235–42. New York: Reinhold, 1935. Silica gel is formed by dehydrating "silicic acid." The elasticity and plasticity of gels produced under acid, neutral, and alkaline conditions are measured by bending tests. For samples of the same age, the elastic properties are similar, but the plasticity and shrinkage increase continuously from the acid to the alkaline gels. Mentions silica esters as binders for powdered silica and calcium carbonate.

1938

King, G., and A. R. Warnes. Improvements relating to the treatment of porous surfaces such as that of stone and to compositions therefor. U.K. Patent Application 6992/38, 5 March 1938; and U.K. Patent 513,366, issued 11 October 1939. The treatment is a solution consisting of or containing "silicic acid ester," or a mixture of "silicic acid esters," or a solution of silica in a nonaqueous solvent(s) or any liquid capable of assuming a permanent gel-like condition in the pores used in conjunction with a fungicide, insecticide, germicide, and so on.

1942

Stone, J. B., and A. J. Teplitz. Earth consolidation. U.S. Patent 2,281,810, issued 5 May 1942. Claims alkylsilicates for use in earth consolidation, particularly for use with oil wells.

1945

Robinson, S. R. An introduction to organosilicon chemistry. *Scientific Journal of the Royal College of Science* 15:24–39. Summary of organosilicon chemistry. Notes use of "ethyl orthosilicates" for preservation of stone.

Shaw, C., and J. E. Hackford. Development in the application of silicic esters. *Industrial Chemist* 21(March):130–35. Reviews ethyl silicates and their industrial applications. Mentions that problems have been encountered with "inexplicable variations" in the behavior of alkoxysilane solutions. Also reviews the hydrolysis reaction and experimental work and mentions that gel formation is accelerated by alkaline materials but forms porous films. Points out that differences in reaction conditions (temperature, reaction time, pH) will affect the hydrolysis reaction and the final product.

1946

Cogan, H. D., and C. A. Setterstrom. Properties of ethyl silicate. *Chemical and Engineering News* 24(18):2499–2501. Reviews methods of preparing hydrolyzed solutions of ethyl silicate.

1947

Cogan, H. D., and C. A. Setterstrom. Ethyl silicates. *Industrial and Engineering Chemistry* 39(11):1364–68. Reviews physical properties and methods of manufacture of ethyl silicates. Three types of commercial ethyl silicates are available: the monomer (TEOS), ethyl silicate (mainly monomer + some polysilicate), and ethyl silicate 40 (a mixture of ethylpolysilicate with an available silica content of about 40). [Same as 1946 article.]

Emblem, H. G. Organo-silicon compounds in paint media: A review of recent developments. *Paint Manufacture* 17(7):239–40. Mentions modified "silicon esters" as being more stable than silicone-based paints. They may be used as concrete/stone preservatives by diluting the ester with a solvent such as white spirits. The setting takes place through the action of atmospheric moisture.

Penn, W. S. Silicic ester plastics. *Australian Plastics* 2:36–43. Describes commercially available "hardeners," Kexacrete, for porous surfaces.

Wagner, H. *Taschenbuch des Chemischen Bautenschutzes.* 3d ed. Stuttgart: Wissenschaftliche Verlagsgesellschaft, 1947. Briefly mentions "silicic acid ester" for the consolidation of deteriorated stone.

1948

Emblem, H. G. Recent developments in silicon ester paints. *Paint Manufacture* 18:359–60. Reviews use of "silicon esters" for masonry preservation.

Emblem, H. G. Silicon ester paints. *Paint Technology* 13(152):309–11. Gives recipes for two paint bases that may also be used, without pigmentation, as stone binders. Solution A is ethyl silicate + alcohol (industrial methylated spirits), and solution B ethyl silicate, alcohol, distilled water with 1% (v/v) of a 0.6% (v/v) HCl solution. The pigments added to the paint bases should be nonbasic. Films formed from the silicon ester solutions are porous and will not protect the substrate from extreme humidity. Dibutyl phthalate found to be the most satisfactory plasticizer for the rigid films that form. Finally, stone hardener should consist of ethyl silicate, alcohol, and an inert solvent (i.e., white spirits) applied in repeated treatments to the surface.

Jullander, I. Treatment of paper with silicon esters. *Nature* 162:300–301. Unsized papers exposed to "silicon ester" vapors were made water impermeable and greater in wet strength.

1949

Thomson-Houston Co. Improvements in and relating to methods of preparing alkyl-orthosilicates. U.K. Patent Application 16,234/47, 19 June 1947; and U.K. Patent 629,138, issued 13 September 1949. Describes a method for the preparation of "silicic acid esters" of organo hydroxy compounds (aliphatic esters of ortho-silicic acid) by reacting alcohol with magnesium silicide.

1950

Aelion, R., A. Loebel, and F. Eirich. Hydrolysis of ethyl silicate. *Journal of the American Chemical Society* 72(12):5705–12. Fundamental study of acid and base catalyzed reactions of TEOS in water and ethanol solutions.

1955

Liberti, S. Consolidamento dei materiali da costruzione di monumenti antichi. *Bolletino dell'Istituto Centrale del Restauro* 21–22:43–70. Reports on testing of ethyl silicate for consolidation of brickwork.

1956

Plenderleith, H. J. *The Conservation of Antiquities and Works of Art.* London: Oxford University Press, 1956. Reports that "for sandstone and siliceous limestone of large dimensions which are kept indoors, a most successful strengthening agent is silicon ester."

Wagner, H. *Taschenbuch des Chemischen Bautenschutzes.* 4th rev. ed. Stuttgart: Wissenschaftliche Verlagsgesellschaft, 1956. Notes that "silicic acid esters" are being replaced by silicones for use as stone preservatives in Germany.

1957

Shore, B. C. G. *Stones of Britain.* London: Leonard Hill Books, 1957. TEOS in an alcohol solution is applied to stone and examined under microscope. "The silica tends to form connecting bars between the ooliths of a limestone or the grains of a sandstone." Mentions a silicon ester used to consolidate a chalk structure/object in 1938.

Smith, D. Cleaning inscriptions and sculptures in sandstone. *Museums Journal* 57:215–19. Stones were consolidated with "colloidal ethyl silicate" after cleaning.

1959

Anon. Stone preservatives. *Building Research Station Digest* 128:1–2. Suggests use of "silicon esters" for use on objects that could be dismantled and submerged in solution.

Blasej, J., J. Doubrava, and J. Rathousky. Pouziti organokremicitych látek pro konservaci a restauraci cásti piskovcoveho zábradli letohrádku v Královske zahrade. *Zprávy Pamatkove Péče* 19:69–80. Describes the use of ethyl silicate and methyltrialkoxysilane mixtures for stone preservation.

General Electric Co. Improvements relating to organosilicon compounds. U.K. Patent Application 11,174/57, 5 April 1957. U.K. Patent 813,520, issued 21 May 1959. A new method of making water-soluble organosilicon compounds useful in imparting water repellency by partially condensing an alkyltrialkoxysilane in ethylene glycol. Alkyl radicals are methyl or ethyl groups, and a mixture of alkoxysilanes is preferred.

1963

Sneyers, R. V. 2e rapport sur l'étude des matériaux pierreux. Report presented at ICOM meeting, Leningrad and Moscow, 16–23 September 1963. Mimeograph. Reports on treatment of calcareous and dolomitic stones with ethyl silicates.

1967

Lerner, R. W., and A. R. Anderson. Silicon-containing water repellent compositions. U.S. Patent 3,310,417, issued 21 May 1967. Describes a mixture of an alkyltrimethoxysilane and a tetraalkoxysilane and water for production of a water-repellent film.

1968

Munnikendam, R. A. Preliminary notes on the consolidation of porous building materials by impregnation with monomers. *Studies in Conservation* 12(4):158–62. Briefly mentions "silicon ester" as a consolidant for sandstone. The inorganic silica is said to be deposited from solution in the pores of the stone.

Sneyers, R. V., and P. J. de Henau. The conservation of stone. In *The Conservation of Cultural Property*, 230–31. Paris: UNESCO Press, 1968. Discusses a product called NuBold for sandstones and siliceous limestones made mainly from ethyl silicate in alcohol. The stone should be dry before treatment, and the number of applications is governed by the permeability of the stone. Work should be discontinued at the first sign of "a white veil on the surface."

1968–1969

Bruno, A., G. Chiari, G. Giullini, C. Trossarelli, and G. Bultinck. Contributions to the study of the preservation of mud-brick structures. *Mesopotamia* 3–4:441–73. Preliminary results of field trials showed ethyl silicate to be best consolidant of earthen structures.

Bultinck, G. A note on the preservation and consolidation of mud-brick work. *Mesopotamia* 3–4:471–73. Experimented with the consolidation of mud-brick with an ethyl silicate solution: 1 volume of HCl (37%) and 34 volumes of denatured alcohol were poured into 65 volumes of ethyl silicate with constant stirring. The solution could be kept stored in a closed container for a week before applying to the mud-brick. The consolidated mud-brick was strong, not darkened, and highly resistant to water with open pores.

1969

Schmidt-Thomsen, K. Zum Problem der Steinzerstorung und -konservierung. *Deutsche Kunst- und Denkmalpflege* 27:11–23. Reports on sandstone consolidation experiments using a solution of ethyl silicate and methyltriethoxysilane, ethyl alcohol, and hydrochloric acid.

1970

Bultinck, G. Preliminary report on the study of the conservation of excavated mud-brick structures. Paper presented at the fourth meeting of ICOMOS, Colloque sur l'alteration des pierres, Brussels, December 1970. Mimeograph. Reports that ethyl silicates (Silester OS, Monsanto, and Ethyl Silicate 40, Union Carbide) formed a hard surface layer without adhesion to the underlying mud-brick.

1971

Kaas, R. L., and J. L. Kardos. The interaction of alkoxysilane coupling agents with silica surfaces. *Polymer Engineering and Science* 11(1):11–18. Studies the interactions of several alkoxysilanes with silica surfaces by infrared spectroscopy and finds evidence of primary chemical bonding.

Munnikendam, R. A. Acrylic monomer systems for stone impregnation. In *La conservazione delle sculture all'aperto*, ed. R. Rossi-Manaresi and E. Riccòmini, 221–28. Bologna: Edizione Alfa, 1971. Examines acrylic resin consolidants but mentions that "methacrylic esters with trimethoxysilane end groups" may be used to increase adhesion of the acrylic resin to inorganic surfaces.

Plenderleith, H. J., and A. E. A. Werner. *The Conservation of Antiquities and Works of Art.* 2d ed. London: Oxford University Press, 1971. The application of ethyl silicate solutions (mentions NuBold) as stone consolidants are briefly reviewed. The stone should be dry before treatment. The consolidant should be sprayed on in three or more applications, but between three and seven applications a "permanent milky surface" could appear, at which point the treatments should stop. Good results were achieved on sandstone, but the products were not suitable for limestone and lavas.

Riederer, J. Die Dokumentation von Steinkonservierungsarbeiten mit Hilfe von Lochkarten. *Arbeitsblatter für Restauratoren* 4:1–6. Describes filing system to cross-reference German monuments and their treatments and includes references to Goldschmidt's sandstone consolidant.

Riederer, J. Stone preservation in Germany. In *Preprints of the New York Conference on Conservation of Stone and Wooden Objects, 7–13 June 1970,* 2d ed., vol. 1, ed. G. Thomson, 125–34. London: IIC, 1971. A general review. Notes that "organic silica compounds," particularly silicate esters, need to be applied by experienced personnel. More research needs to be done to evaluate alkoxysilanes but recommends silicate esters for saving crumbling smaller objects only if "there is no other process which can help."

Shore, B. C. G. Conservation of valuable buildings. In *The Monument for the Man,* 264–67. Padua: ICOMOS and Marsilio, 1971. Reviews early alkoxysilane research.

Torraca, G. An international project for the study of mud-brick preservation. In *Preprints of the New York Conference on Conservation of Stone and Wooden Objects, 7–13 June 1970,* 2d ed., vol. 1, ed. G. Thomson, 47–57. London: IIC, 1971. Detailed report on field testing of ethyl silicate (Silester OS, Monsanto) in Iraq.

1971–1972
Bultinck, G. De conservatie van ruines in ongebakken klei: Een samenvattend overzicht. *Bulletin de l'Institut Royal du Patrimoine Artistique* 13:131–38. Report on mud-brick treatment project with ethyl silicate (Silester OS, Monsanto) in Iraq.

1972
Anon. Building preservatives fall on stony ground. *New Scientist,* 31 August 1972, 437. Reports the failures of some stone consolidants. Based on Building Research Establishment report by Clarke and Ashurst (1972).

Chvatal, T. Moderne Chemie hilft den Bauwerken. *Maltechnik-Restauro* 78:131–38. Reviews chemistry of alkoxysilanes used for the consolidation of stone. Results of mechanical testing of consolidated samples are discussed. Focus is on catalyzed systems.

Chvatal, T. Moderne Chemie hilft den Bauwerken. *Steinmetz und Bildhauer* 7:147–54. Same as above.

Clarke, B. L., and J. Ashurst. Stone preservation experiments. London: H.M. Stationery Office, 1972. Reports on a number of consolidants tested in the field. Notes no overall beneficial effects by any, including a "silicon ester."

Cuenca, R. S. Alteración y consolidación de la piedra en los monumentos. In *De Re Restauratoria,* vol. 1, 155–61. Barcelona: Escuela Tecnica y Superior de Arquitectura de Barcelona 1972. States that alcohol, ethyl silicate, and hydrochloric or phosphoric acid solutions are effective stone consolidants.

Hempel, K., and A. Moncrieff. Summary of work on marble conservation at the Victoria and Albert Museum Conservation Department up to August 1971. In *The*

Treatment of Stone, ed. R. Rossi-Manaresi and G. Torraca, 165–81. Bologna: Centro per la Conservazione delle Sculture all'Aperto, 1972. Rhone-Poulenc X54-802 (MTMOS) was found to vary from batch to batch: the color varied from colorless to pale yellow, and the curing time ranged from a few hours to several days. Decayed marble was treated with the X54-802 and exposed to the London atmosphere. The treated samples showed no sign of breakdown, whereas the untreated sample had deteriorated "to a pile of dust." X54-802 did not cure well on Verona marble (it took a long time to gel). The addition of a small amount of potassium hydroxide was found to improve the cure. The X54-802 was tested with a variety of powdered substances to be used as a filler. It was found that chalk, kaolin, sepiolite, and aluminum oxide caused the alkoxysilane to cure slowly. Mixtures of the X54-802 with powdered Verona and Istrian stone caused an even slower cure (i.e., longer time to gelation).

Lewin, S. Z. Recent experience with chemical techniques of stone preservation. In *The Treatment of Stone,* ed. R. Rossi-Manaresi and G. Torraca, 139–44. Bologna: Centro per la Conservazione delle Sculture all'Aperto, 1972. Reports on successful results with alkoxysilanes for porous sandstone, mortar, and concrete. Mentions a new approach based on the "controlled hydrolysis of a tetraalkoxysilane" on decayed sandstone, lava stone, tuff, dolomite, and concrete.

Luckat, S. Investigations concerning the protection against air pollutants of objects of natural stone. *Staub-Reinhaltung der Luft* 32(5):30–33. Reports on the performance of consolidated stone in the sodium sulfate crystallization test. Tegovakon products were reported to give good results.

Munnikendam, R. A. The combination of low viscosity epoxy resins and silicon esters for the consolidation of stone. In *The Treatment of Stone,* ed. R. Rossi-Manaresi and G. Torraca, 197–200. Bologna: Centro per la Conservazione delle Sculture all'Aperto, 1972. Used ethyl silicates or methyl silicates as diluents for epoxy consolidants. 50:50 mixtures of epoxy-alkoxysilane gave unexpectedly good results. The viscosity of the mixture was low, and the material had good color, mechanical properties, and water resistance. States that a few percent of the "silicon ester" may be replaced by MTMOS to impart water repellency.

Riederer, J. The conservation of German stone monuments. In *The Treatment of Stone,* ed. R. Rossi-Manaresi and G. Torraca, 105–24. Bologna: Centro per la Conservazione delle Sculture all'Aperto, 1972. A general review of stone conservation in Germany. Reports the growing use of "silicic esters" as sandstone consolidants. Refers to "Sandstein Festiger" as a two-component system containing ethyl silicate, MTMOS, and HCl as the catalyst. Also mentions another product based on silicic esters with a phosphoric acid catalyst, but no name or details are given.

Riederer, J. Schäden an kunstdenkmälern-ihre Vermeidung durch regelmässige Pflege. *Gebaudereiniger-Handwerk* 7:4–14. Recommends "silicic acid ester" stone consolidants.

Riederer, J. Steinkonserviering im Lichte neuerer Erkenntnisse. *Eurafem-Information* 3:1–2. Recommends "silicic acid ester" stone consolidants.

Riederer, J. Stone conservation with silicate esters. Paper presented at the meeting of the ICOM Committee for Conservation, 2–8 October 1972, Madrid. Mimeograph. Reports on the use of alkoxysilane stone consolidants including testing and product information. Lists German monuments successfully treated with the original Goldschmidt product.

Sobolevskii, M. V., G. S. Popeleva, Z. G. Brykova, O. A. Muzovskaya, K. P. Grinevich, and A. S. Denisova. Organosilicon water repellents and their effectiveness in building. *Soviet Plastics,* 34–37. Recommends the incorporation of ethyl silicate 40 in finishing materials for walls.

Stambolov, T., and J. R. J. van Asperen de Boer. *The Deterioration and Conservation of Porous Building Materials in Monuments: A Literature Review.* Rome: ICCROM, 1972. The authors abstract five articles from the ethyl silicate literature based on the work of King, Threlfall, and Wagner.

1973

Aguzzi, F., A. Fiumara, A. Peroni, R. Ponci, V. Riganti, R. Rossetti, F. Soggetti, and F. Veniale. L'arenaria della Basilica di S. Michele in Pavia: Ricerche sull'alterazione e sugli effetti dei trattamenti conservativi. *Atti Societa Italiana di Scienze Naturali* 114(4):403–64. Reports on Lewin's 1971 test treatments with ethyl silicate.

Bauer-Bornemann, U. Restaurierungsarbeiten an der Kirche "Zu unserer Lieben Frau" in Bamberg (1320–1380). *Th. Goldschmidt* 3(4):27. Describes treatment of sandstone with Tegovakon products.

Bosch, E. Use of silicones in conservation of monuments. In *First International Symposium on the Deterioration of Building Stones, 11–16 September 1972,* 21–26. Chambery: Imprimeries Reunies, 1973. Describes new product, Wacker-Chemie VP 1301.

Luckat, S. Untersuchungen zum Schutz von Sachgütern aus Naturstein vor Luftverunreinigungen. *Th. Goldschmidt* 3(4):19–23. Reports on tests with Tegovakon products on sandstone.

Riederer, J. Die Erhaltung historischer Grabdenkmäler. *Steinmetz und Bildhauer* 5:272–75. Reports on treatment of sandstone with "silicon ester."

Riederer, J. Die Erhaltung von Kunstwerken aus Stein in Deutschland. *Maltechnik-Restauro* 79:6–30. Reports on treatment of sandstone with "silicon ester."

Riederer, J. Steinkonservierung am Aphaia-Tempel auf Aegina. *Maltechnik-Restauro* 79:193–99. Reports on testing of consolidants on limestone, including Tegovakon products. Hoffmann's water glass system found to have best results.

Riederer, J. Steinkonservierung in Bayern. *Jahrbuch der bayerischen Denkmalpflege* 28:264–83. Lists twenty-five monuments treated with Goldschmidt's sandstone consolidant. Notes type of stone, date of treatment, and conservator.

Seiler, C. D. Method for waterproofing masonry structures. U.S. Patent 3,772,065, issued 13 November 1973. Claims use of alcohol and alkyltrialkoxysilane solutions to waterproof masonry.

Worch, E. Steinkonservierung. *Th. Goldschmidt* 3(24):24–26. Reports on practical experience with Tegovakon products and states that calcareous stones slow condensation reactions.

1974

Chvatal, T. Die Festigung von Stein: Chemische Grundlagen und praktische Erfahrungen mit Kieselsäureestern. *Arbeitsblatter für Restauratoren* 7:40–51. Transcript of lecture to museum technicians. Reviews physical properties of alkoxysilanes used as stone consolidants. Author rejects acid catalysts due to their reaction with calcareous stones.

Chvatal, T. Vergleichende untersuchung der Wichtigsten Modernen Steinkonservierungs mittel. *Maltechnik-Restauro* 80:87–97. Reports results of a testing program of consolidants on calcareous sandstones.

Hempel, K. F. B. Conservation of external calcareous sculpture. *Stone Industries* 9:14–16. Brief review of the use of methyltrimethoxysilane as a stone consolidant.

Oel, H. J., and H. Marschner. *Physikalisch-chemische Untersuchungen zur Konservierung von Bauplastiken und freistehenden Skulpturen.* Final report of the Institut für Werkstoffwissenschaften III, Erlangen, 1974, 74 pp. Reports on the consolidation of sandstone cubes and their performance in the salt crystallization test.

Plankl, L. H., C. D. Seiler, and H. J. Vahlensieck. Surface protection of porous materials. U.S. Patent 3,819,400, issued 25 June 1974. Method to protect porous masonry involving organo silicon compounds claimed by Dynamit Nobel AG.

Riederer, J. Die Erhaltung Ägyptischer Baudenkmaler. *Maltechnik-Restauro* 80:43–52. "Silicon esters" recommended for the consolidation of sandstone and salt-laden granite.

Riederer, J. Pollution damage to works of art. In *New Concepts in Air Pollution Research,* ed. J.-O. Willums, 73–85. Basel: Birkhauser Verlag, 1974. Mentions silicon esters for use in stone consolidation.

1975

Chiari, G. Conservación de los monumentos arqueológicos en adobe: Peru. UNESCO RLA/047/72-FMR/SHC/OPS/243. Work with ethyl silicate indicates that better -consolidation is achieved without acid catalyst.

Iakachvili, T. V. Problèmes de consolidation et de hydrophobation de la surface d'un massif de roche et des murs intérieurs des cavernes d'un monument du XIIe siècle "Vardzia" (Géorgie). Paper presented at the fourth meeting of the ICOM Committee for Conservation, 13–18 October 1975, Venice, vol. 1, no. 75/5/6. Mimeograph. Reports on treatment of tufaceous caverns with an aqueous solution of silicates.

Luckat, S. Die einwirkung von Luftverunreinigungen auf die Bausubstanz des Kölner Domes. III. *Kölner Dombladt* 40:75–108. Reports on testing of proprietary consolidants on sandstone, trachyte, basalt, and limestone. Recommends certain products for decayed stone and others for undecayed stone. Some products (Tegovakon) were found to give poorer results than untreated stone.

Nestler, H., J. Amort, and L. H. Plankl. Composition for impregnation of masonry having a neutral or acidic reaction surface. U.S. Patent 3,879,206, issued 22 April 1975. Alcohol or hydrocarbon solution of alkyltrialkoxysilane for impregnation of sandstone or carbonated concrete to reduce water absorption claimed for Dynamit Nobel AG.

Price, C. A. The decay and preservation of natural building stone. *Building Research Establishment Current Paper,* CP89/75, 350–53. Watford: Building Research Establishment, 1975. Discusses stone decay and reviews treatments (acrylics, alkoxysilanes, and epoxies).

Stambolov T., and J. R. J. van Asperen de Boer. The deterioration and conservation of porous building materials in monuments: A literature review. Supplement 1975. Paper presented at the fourth meeting of the ICOM Committee for Conservation,

13–18 October 1975, Venice, vol. 1, no. 75/5/8. Mimeograph. Supplement to the 1972 bibliography, adding two articles by Chvatal.

Winkler, E. M. *Stone: Properties, Durability in Man's Environment.* 2d rev. ed. New York: Springer-Verlag, 1975. Refers to "organic silicates" as the best consolidants for sandstone in a brief review of stone consolidants.

1976

Arnold, L., D. Honeyborne, and C. A. Price. Conservation of natural stone. *Chemistry and Industry* 17(8):345–47. TEOS/MTEOS and water solutions gave some increase in mechanical strength to limestone.

Arnold, L., and C. A. Price. The laboratory assessment of stone preservatives. In *The Conservation of Stone I,* ed. R. Rossi-Manaresi, 695–704. Bologna: Centro per la Conservazione delle Sculture all'Aperto, 1976. TEOS and MTMOS were mixed with the stoichiometric amounts of water and a trace of acid and applied to samples of Richemont stone. The tensile strengths were determined, and the alkoxysilanes performed better than the methylmethacrylate and about the same as the barium hydroxide treatment. All performed worse than those samples consolidated with molten paraffin wax. It was found that the MTMOS solutions did not immobilize salts.

Bosch, E., M. Roth, and K. Gogolok. Binder composition for inorganic compounds. German Patent 2,318,494, 31 October 1974; U.S. Patent 3,955,988, issued 11 May 1976. Patent for a stone "binder" of alkylsilicates and organic solvent(s) and the metallic salt of a carboxylic catalyst and an organoalkylsilane (optional). This is the patent for Wacker H and Wacker OH.

Chvatal, T. Zum Hydrophobieren von Steinmaterial. *Arbeitsblatter für Restauratoren* 9:64–82. Lecture mainly on water repellents but mentions Wacker H and MONUMENTIQUE (Bau-Chemie).

Hempel, K. F. B. An improved method for vacuum consolidation of decayed stone sculpture. In *Second International Symposium on the Deterioration and of Building Stones,* ed. T. Skoulikidis, 163–66. Athens: Ministry of Culture and Science of Greece, 1976. Discusses treatment of stone sculptures with "silane monomer" (Rhone-Poulenc X54-802).

Hempel, K., and A. Moncrieff. Report on work since last meeting in Bologna, October 1971. In *The Conservation of Stone I,* ed. R. Rossi-Manaresi, 319–39. Bologna: Centro per la Conservazione delle Sculture all'Aperto, 1976. Deteriorated Carrara marble was treated with a "high density silicone monomer resin" and exposed, on a roof, to the London atmosphere for six years. At the end of this time, the untreated sample had completely decomposed, while the treated sample appeared unaltered. A typical consolidation would be applied in several steps. First, the object was exposed to a solvent. The X54-802 was then mixed with 50% solvent and approx. 5% water (amount of water was increased or decreased by a few percent depending on the RH). The following steps consist of applying more consolidant but in greater concentration, that is, a 2–1 mixture followed by a 5–1 mixture. When used as a fill material, marble was found to increase the curing time. Tested a number of consolidants on Carrara marble and limestone: Tegovakon (2 pack), Wacker VP (later Wacker H), EP5850 10% HCl soln, X1-9010, X54-802. All were performing well after 15 months' exposure on the roof. The X54-802 was the only one found to positively consolidate powdered marble.

Hoke, E. Microprobe investigations of incrusted as well as cleaned marble specimens. In *Second International Symposium on the Deterioration of Building Stones*, ed. T. Skoulikidis, 119–26. Athens: Ministry of Culture and Science of Greece, 1976. Results show uneven penetration of "silicon esters" into marble. Weathering tests conducted showed films to be stable, but weathering processes continued on marble.

Moncrieff, A. The treatment of deteriorating stone with silicone resins: Interim report. *Studies in Conservation* 21(4):179–91. Reports on laboratory experiments of Rhone-Poulenc X54-802. Reviews chemistry and testing of various solution formulations. Hypothesizes that product encapsulated salts in stone.

Moncrieff, A. The treatment of deteriorating stone with synthetic resins: A further report. In *Second International Symposium on the Deterioration of Building Stones*, ed. T. Skoulikidis, 167–69. Athens: Ministry of Culture and Science of Greece, 1976. Reports on consolidation work at the Victoria and Albert Museum. Best results on limestone and marble with Rhone Poulenc X54-802 (MTMOS), which consolidates "crumbly" stone. Performed well on marble exposed to the London environment for seven years. Wacker H was found to consolidate crumbly stone if applied several times and allowed to cure, between applications, at high relative humidity.

Riederer, J. Further progress in German stone conservation. In *The Conservation of Stone I*, ed. R. Rossi-Manaresi, 369–74. Bologna: Centro per la Conservazione delle Sculture all'Aperto, 1976. General overview of the increasing use of inorganic solutions for stone conservation in Germany. The use of ethyl silicates has dramatically increased whereas polymers/resins and soluble (alkali) silicates have decreased. The advantages of ethyl silicate are listed as (1) having good penetration (approx. 4 cm in ordinary sandstone) and (2) the cracks that occur as the gel ages are amenable to retreatment. States that porous stones are successfully treated by ethyl silicate consolidants.

Rossi-Manaresi, R. Treatments for sandstone consolidation. In *The Conservation of Stone I*, ed. R. Rossi-Manaresi, 547–71. Bologna: Centro per la Conservazione delle Sculture all'Aperto, 1976. Tested several consolidants (Rhodorsil X54-802, XR-893, ACRYLOID B72 + Dri Film 104, Wacker H, Tegovakon, epoxy resin and polyurethane resin). The treatments that were applied in situ did not penetrate as deeply as when applied under laboratory conditions.

Stambolov, T., and J. R. J. van Asperen de Boer. *The Deterioration and Conservation of Porous Building Materials in Monuments: A Review of the Literature*. 2d rev. ed. Rome: ICCROM, 1976. Combines 1972 and 1975 contributions.

Taralon, J., C. Jaton, and G. Orial. Etat des recherches effectuées en France sur les hydrofuges. In *The Conservation of Stone I*, ed. R. Rossi-Manaresi, 455–76. Bologna: Centro per la Conservazione delle Sculture all'Aperto, 1976. Recommends the use of "polymère de l'acide silicique" for use on tufa and limestone.

Torraca, G. Brick, adobe, stone and architectural ceramics: Deterioration processes and conservation practices. In *Preservation and Conservation: Principles and Practices*, ed. S. Timmons, 143–65. Washington, D.C.: Preservation Press, 1976. Mentions ethyl silicate and methytriethoxysilane for the consolidation of stone.

Weber, H. Stone renovation and consolidation using silicones and silicic esters. In *The Conservation of Stone I*, ed. R. Rossi-Manaresi, 375–85. Bologna: Centro per la Conservazione delle Sculture all'Aperto, 1976. Claims that strengtheners H and OH are efficient for sandstone, limestone, ceramic building materials (terracotta), and sometimes marble. The best results are on highly porous and absorbent stones. The "penetration power" is an advantage for these materials, and using enough material is important to good consolidation (2–20 l/m²). Marble consolidated with "silicon ester" gave modest increases in compressive strength (20–50 kg/cm²) and good durability when subjected to the salt crystallization test.

Wihr, R. Deep impregnation for effective stone protection. In *The Conservation of Stone I,* ed. R. Rossi-Manaresi, 317–18. Bologna: Centro per la Conservazione delle Sculture all'Aperto, 1976. Documents the application of OH on a sandstone statue by setting it in a shallow tub, surrounded by a spray system and enclosed in a cabinet of polyethylene sheeting. This technique was adapted for buildings by the use of spray jets and collecting and recycling the runoff. Penetration of up to 25 cm is achieved.

1977

Anon. John Ashurst: Irreversible chemical processes, "the way of the world." *Architects' Journal,* 23 November 1977, 1028. Reports on the controversy over Brethane treatment of Malmsbury Abbey under the direction of John Ashurst.

Caroe, M. B. Wells Cathedral, The West Front Conservation Programme. Interim report on aims and techniques, June 1977. Color changes noted after alkoxysilane treatment.

Klemm, D. D., R. Snethlage, and B. Graf. Bericht über die Steinrestaurierung am Bayer-Tor in Landsberg am Lech. *Maltechnik-Restauro* 83:242–52. Reports treatment of calcareous sandstone with Wacker OH.

Marsh, P. Breathing new life into the statues of Wells. *New Scientist,* 22–29 December 1977, 754–56. Describes planned treatment of limestone statues with alkoxysilanes (Brethane and X54-802).

Moncrieff, A., and K. F. B. Hempel. Conservation of sculptural stonework: Virgin and Child on S. Maria dei Miracoli and the loggetta of the Campanile, Venice. *Studies in Conservation* 22(1):1–11. The treatment tested was 100 g Rhone Poulenc X54-802 (MTMOS), 10 g 2-propanol, 4 g deionized water, 100 g 2-ethoxyethanol. Some test pieces taken from the objects had been coated with "grease" to keep the birds away, and on these the treatment did not properly cure. For the consolidation, the stone first was exposed to 2-ethoxyethanol, and then the consolidant was applied in several coats. Fills were also prepared from the alkoxysilane and mineral powders (marble, glass, quartz) with inconsistent results.

Riederer, J. Die Probleme der Steinkonservierung in Ceylon. *Maltechnik-Restauro* 83:41–50. Recommends the use of alkoxysilanes for consolidation of seriously deteriorated crystalline limestone. Alkoxysilane products reported to inhibit water penetration and plant growth.

Rossi-Doria, P., M. Laurenzi Tabasso, and G. Torraca. *Nota sui trattamenti conservativi dei manufatti lapidei: Atti ICR— laboratorio prove sui materiali.* Rome: Istituto Centrale per il Restauro, 1977. Reviews consolidation treatments and discusses inter alia ethyl silicates, alkyltrialkoxysilanes, and mixtures of ethyl silicates and alkyltrialkoxysilanes.

Sleater, G. A. Stone preservatives: Methods of laboratory testing and preliminary performance criteria. In *NBS Technical Note 941.* Washington, D.C.: U.S. Government Printing Office, 1977. Reports on testing of ethyl silicate solutions. Depth of penetration reported for ethyl silicate was low.

Weber, H. Ursachen und Behandlung der Steinverwitterung. *Maltechnik-Restauro* 83:73–89. Reports on use and testing of Wacker products. Repeated applications of a "silicon ester" product over one to two weeks said to be an effective method.

Worch, E. Gedanken eines Praktikers zur Steinkonservierung. *Maltechnik-Restauro* 83:253–55. Discusses performance differences of alkoxysilane-based consolidants, particularly Tegovakon products, in the laboratory and the field.

1978

Arnold, L. The preservation of stone by impregnation with silanes. *Newsletter of the Council for Places of Worship*, no. 24:4–6. Describes treatment of statue 117 at Wells Cathedral with Brethane and indicates that it should not be used if the object is severely contaminated with salt.

Cormerois, P. Traitements préventif et curatif des structures. In *Preprints for the International Symposium: Deterioration and Protection of Stone Monuments, 5–9 June 1978*, vol. 2, no. 6.5. Paris: RILEM and UNESCO, 1978. Suggests mixing X54-802 with Rhodorsil 1330 C (a mixture of methyl(phenyl)polysiloxane, epoxy resin, linoleic acid) and an "aminoorganosilane" in order to improve film-forming capabilities.

Kotlik, P., and J. Zelinger. Resistance of sandstone hardened by polymer agents against the influence of aqueous salts solutions. In *Preprints for the International Symposium: Deterioration and Protection of Stone Monuments, 5–9 June 1978*, vol. 2, no. 6.9. Paris: RILEM and UNESCO, 1978. Porous sandstones consolidated with Wacker OH, Tegovakon, and Wacker 190 S were evaluated with sodium sulfate crystallization test.

Marschner, H. Application of salt crystallisation test to impregnated stones. In *Preprints for the International Symposium: Deterioration and Protection of Stone Monuments, 5-9 June 1978*, vol. 1, no. 3.4. Paris: RILEM and UNESCO, 1978. Uses a modified version of DIN 52111 for testing 45 consolidants on stone.

Penkala, B., K. Krajewski, and R. Krzywoblocka-Laurow. Study of the effectiveness of the methods for stones preservation. In *Preprints for the International Symposium: Deterioration and Protection of Stone Monuments, 5-9 June 1978*, vol. 2, no. 6.1. Paris: RILEM and UNESCO, 1978. Reports on laboratory testing of three types of "silico-organic compounds."

Price, C. A. Verfahren und flussiges Mittel zur behandlung von porosen anorganischen materialien. German Patent 2,733,686, issued 2 February 1978. Claims the use of a three-part mixture for stone consolidation, assumed to be Brethane.

Riederer, J. Recent advances in stone conservation in Germany. In *Decay and Preservation of Stone*. Engineering Geology Case Histories, vol. 11, ed. E. M. Winkler, 89–93. Boulder, Colo.: Geological Society of America, 1978. A general review of stone conservation. The advantages of ethyl silicate are depth of impregnation retains appearance of stone; can be mixed with water repellents; easy application; can be re-treated. Documents treatments still lasting after fifteen years.

Riederer, J. Steinkonservierung in Bayern. *Jahrbuch der Bayerischen Denkmalpflege* 30:223–37. Reports on the treatment of 46 monuments during the years 1971–74 with a Goldschmidt product and Wacker VP 1301.

Snethlage, R., and D. D. Klemm. Scanning electron microscope investigations on impregnated sandstones. In *Preprints for the International Symposium: Deterioration and Protection of Stone Monuments, 5–9 June 1978*, vol. 2, no. 5.7. Paris: RILEM and UNESCO, 1978. Uses SEM to study coating and adhesion of Wacker OH, H, and

190S in sandstone. OH films had drying cracks but filled spaces between pores with wedges of polymer. Wacker H reportedly deposited in two phases.

Weber, H. Stellungnahme. *Maltechnik-Restauro* 84:65. Comments on discrepancies between laboratory and in situ results as reported by Worch (1977).

Wihr, R. The use of aethyl-silicate and acrylic-monomers in stone preservation. In *Preprints for the International Symposium: Deterioration and Protection of Stone Monuments, 5–9 June 1978*, vol. 3, no. 7.12. Paris: RILEM and UNESCO, 1978. Reports good results with ethyl silicate on certain types of stone.

1979

Berti, P. Prove di laboratorio su elementi di mattoni trattati con resine siliconiche. In *Il Mattone di Venezia: Stato delle conoscenze tecnico-scientifiche: Atti del convegno presso Fondazione Cini, 22–23 ottobre 1979*, 439–46. Venice: Laboratorio per lo Studio della Dinamica delle Grandi Masse del CNR e dell'Universita di Venezia, 1979. Salt crystallization tests carried out on brick, concrete, and sandstone consolidated with an alkyltrimethoxysilane solution or a silicic acid ester/methyltriethoxysilane solution.

Lewin, S. Z., and A. E. Charola. The physical chemistry of deteriorated brick and its impregnation technique. In *Il Mattone di Venezia: Stato delle conoscenze tecnico-scientifiche: Atti del convegno presso Fondazione Cini, 22–23 ottobre 1979*, 189–214. Venice: Laboratorio per lo Studio della Dinamica della Grandi Masse del CNR e dell'Universita di Venezia, 1979. Reviews chemistry and presents surface-reacting molecules, such as alkoxysilanes, as the materials offering the most potential.

Roedder, K. M. Zusammensetzung und Wirkungsweise von Silan-Bautenschutzmitteln. In *Kolloquium über Steinkonservierung, Munster, Sept. 1978*, 189–207. Munich: Das Landesamt, 1979. General review of alkoxysilanes. States that ethanol solutions gave the best results in wet stones.

Schmidt-Thomsen, K. Erfahrungen mit konservierungsmittel nach einem Jahnzehnt. In *Kolloquium über Steinkonservierung: Munster, Sept. 1978*, 143–150. Munich: Das Landesamt, 1979. Reviews ethyl silicate stone consolidation products from the 1960s.

Wihr, R. Kieselsäurester und Methylmethacrylate: Zwei wichtige Steinkonservierungs mittel. In *Kolloquium über Steinkonservierung: Munster, Sept. 1978*, 151–58. Munich: Das Landesamt, 1979. General discussion of techniques and treatments for stone conservation.

Worch, E. Gedanken eines Praktikers zur Steinkonservierung. In *Kolloquium über Steinkonservierung: Munster, Sept. 1978*, 159–66. Munich: Das Landesamt, 1979. General discussion of techniques and treatments for stone conservation. Mentions the flaking of silicate/silicone treated layers.

1980

Chiari, G. Treatment of adobe friezes in Peru. In *Third International Symposium on Mudbrick (Adobe) Preservation*, ed. O. Üstünkök and E. Madran, 39–45. Ankara, Turkey: ICOM-ICOMOS, 1980. Better consolidation achieved with ethyl silicate *without* acid catalyst.

Clifton, J. R. *Stone Consolidating Materials—A Status Report.* NBS Technical Note No. 1118. Washington, D.C.: U.S. Government Printing Office. Twenty-one articles are cited in this literature review.

Koller, M., I. Hammer, H. Paschinger, and M. Ranacher. The abbey church at Melk: Examination and Conservation. In *Conservation within Historic Buildings,* ed. N. Brommelle, G. Thomson, and P. Smith, 101–12. London: IIC, 1980. Recommends treating wall paintings damaged by soluble salts with "methyl silicate esters."

Larson, J. The conservation of stone sculptures in historic buildings. In *Conservation within Historic Buildings,* ed. N. Brommelle, G. Thomson, and P. Smith, 132–38. London: IIC, 1980. Reports on the use of methyltrimethoxysilane, an "acrylic silane" (Raccanello E0057) and a mixture of the two. MTMOS did not encapsulate salts.

Rossi-Manaresi, R., and G. Chiari. Effectiveness of conservation treatments of a volcanic tuff very similar to adobe. In *Third International Symposium on Mud-brick (Adobe) Preservation,* ed. O. Üstünkök and E. Madran, 29–38. Ankara, Turkey: ICOM-ICOMOS, 1980. Examines acid catalyzed ethyl silicate, aluminum stearate, and the combination of the two for the consolidation of a volcanic tuff.

Schwartzbaum, P. M., C. S. Silver, and C. Wheatley. The conservation of a Chalcolithic mural painting on mud-brick from the site of Teleilat Ghassul, Jordan. In *Third International Symposium on Mud-brick (Adobe) Preservation,* ed. O. Üstünkök and E. Madran, 177–200. Ankara, Turkey: ICOM-ICOMOS, 1980. Describes treatment of fragments with Wacker H.

Weber, H. *Steinkonservierung.* Berlin: Expert-Verlag, 1980. Reviews stone consolidation treatments.

Wihr, R. The conservation of stone objects in humid interiors. In *Conservation within Historic Buildings,* ed. N. Brommelle, G. Thomson, and P. Smith, 139–41. London: IIC, 1980. Recommends Goldschmidt or Wacker products for consolidation, after desalination.

Wihr, R. *Restaurierung von Steindenkmaelern.* Munich: Verlag Callwey, 1980. Review of the conservation and restoration of stone.

1981

De Witte, E., A. Terfve, and S. Florquin. The identification of commercial water repellents. In *The Conservation of Stone II,* ed. R. Rossi-Manaresi, 577–85. Bologna: Centro per la Conservazione delle Sculture all'Aperto, 1981. Analyzes a number of different commercial products as to their components by XRF and NMR. Silicones (for past treatments) were analyzed by pyrolysis–GC-MS.

Furlan, V., and R. Pancella. Propriétés d'un grès tendre traité avec des silicates d'éthyle et un polymère acrylique. In *The Conservation of Stone II,* ed. R. Rossi-Manaresi, 645–63. Bologna: Centro per la Conservazione delle Sculture all'Aperto, 1981. Studies the properties of sandstone consolidated with ethyl silicates or poly(methyl-methacrylate).

Grissom, C. A., and N. R. Weiss. *Alkoxysilanes in the Conservation of Art and Architecture: 1861–1981.* In *Art and Archaeology Technical Abstracts* 18, no. 1, Supplement, 1981, 150–202. Annotated bibliography on alkoxysilane stone consolidants.

Leary, E. Report on the Brethane treatment of the Great West Doorway at York Minster. *Building Research Establishment Note,* 66/81, Watford: Building Research Establishment, 1981. Discusses the consolidation of limestone with Brethane and indicates an adjustment in the catalyst concentration due to cold weather during the application.

Price, C. A. Brethane stone preservative. *Building Research Establishment Current Paper* CP 1/81, 1–9. Watford: Building Research Establishment, 1981. Brethane is a three-component product mixed immediately before using and is said to be more controllable than other alkoxysilane products. Second application with Brethane causes temporary softening and swelling of the first treatment with no long-term damage. Does not perform well on stone that is heavily contaminated with sodium chloride. Not recommended for stone with rising damp.

Price, C. A. Stone treatment. U.K. Patent 1,588,963, issued 7 May 1981. Claims for Brethane.

Rossi-Manaresi, R. Effectiveness of conservation treatments for the sandstone of monuments in Bologna. In *The Conservation of Stone II,* ed. R. Rossi-Manaresi, 665–88. Bologna: Centro per la Conservazione delle Sculture all'Aperto, 1981. A calcitic sandstone was treated with Rhodorsil X54-802, Rhodorsil XR-893, Wacker H, and Tegovakon T(?), and the Bologna Cocktail. Used Japanese tissue in preconsolidation phase to hold flakes and scales in place.

Roth, M. Paints and impregnants for natural stone. *Bautenschutz und Bausanierung* 4(2):61–63. Discusses silicate-based paints and consolidants.

Sobkowiak, D. Studies on the use of silicoorganic preparations made in Czechoslovakia: Studies of the properties of Silgel IHM and Silgel IEM preparations. *Chemia w Konserwacji Zabytkow, Informator PKZ,* 1981, 113–22. Compares the Silgel products with Wacker H and OH on limestone. All products were found to increase mechanical properties.

Torraca, G. *Porous Building Materials: Materials Science for Architectural Conservation.* Rome: ICCROM, 1981. Reviews some alkoxysilane chemistry. Recommends ethyl silicates for mud-brick.

Twilley, J. Fabrication, deterioration, and stabilization of the Watts Towers: An interim report. In *Preprints of Papers Presented at the Ninth Annual Meeting: Philadelphia, Pennsylvania, 27–31 May 1981,* 182–90. Alkoxysilane consolidation treatment proposed.

Weber, H. Natural stone preservation: The silicic acid ester process. *Bautenschutz und Bausanierung* 4(2):52–55. Reviews the use of ethyl silicate with a catalyst in alcohol or ketone solvents for stone conservation.

Wheeler, G. A short laboratory investigation of a stone consolidant similar to Brethane. *Building Research Establishment Note,* No. 35/81, Watford: Building Research Establishment, 1981. Study of an Austrian alkoxysilane (MTMOS)-based consolidant. The effect of catalyst concentration on gel time, viscosity, and resulting xerogel is reported.

1982

Cavaletti, R., L. Marchesini, and G. Strazzabosco. Tecnologie di consolidamento e di restauro figurative di sculture in pietra tenera dei Colli Berici. In *Third International Congress on the Deterioration and Preservation of Stone Objects, Venezia, 24–27 Ottobre 1979,* ed. B. Badan, 453–60. Padova: Università degli Studi di Padova,

Istituto di Chimica Industriale, 1982. A carbonate-bearing rock with clay inclusion layers is consolidated with alkoxysilanes.

Clifton, J. R., and G. Frohnsdorff. Stone consolidating materials: A status report. In *Conservation of Historic Stone Buildings and Monuments,* ed. N. S. Baer, 287–311. Washington, D.C.: National Academy of Sciences, 1982. Literature review similar to Clifton 1980.

Domaslowski, W. *La Conservation Préventive de la Pierre.* Musées et Monuments, no. 18. Paris: UNESCO, 1982. Summary of stone conservation prepared by the Institute for the Conservation of Cultural Property of the Nicholas Copernicus University in Torun, Poland. Contains an appendix listing synthetic materials and their application to stone conservation.

Feilden, B. M. *Conservation of Historic Buildings.* London: Butterworth, 1982. Mentions ethyl silicates for surface consolidation of adobe.

Fiumara, A., V. Riganti, F. Veniale, and U. Zezza. Sui trattamenti conservativi della pietra d'Angera. In *Third International Congress on the Deterioration and Preservation of Stone Objects, Venezia, 24–27 Ottobre 1979,* ed. B. Badan, 339–56. Padova: Università degli Studi di Padova, Istituto di Chimica Industriale, 1982. Reports on testing of ethyl silicate/alkyltrialkoxysilane solution on dolomitic stone.

Hošek, J., and J. Šrámek. Surface consolidation of sculptures made of gaize. In *Third International Congress on the Deterioration and Preservation of Stone Objects, Venezia, 24–27 Ottobre 1979,* ed. B. Badan, 333–37. Padova: Università degli Studi di Padova, Istituto di Chimica Industriale, 1982. Reports on the treatment of gaize (a marly limestone) with alkylalkoxysilanes (SILGEL).

Rossi-Manaresi, R., G. Alessandrini, S. Fuzzi, and R. Peruzzi. Assessment of the effectiveness of some preservatives for marble and limestones. In *Third International Congress on the Deterioration and Preservation of Stone Objects, Venezia, 24–27 Ottobre 1979,* ed. B. Badan, 357–76. Padova: Università degli Studi di Padova, Istituto di Chimica Industriale, 1982. Tested three treatments of compact limestones. Consolidants tested were Raccanello 0073 (acrylic polymer–silicone resin), Incoloro per marmi 14.21 (acrylic polymer) in toluene/xylenes, and the Bologna Cocktail. Physical characteristics (bulk density, capillarity, water absorption, porosity) were noted, and SEM examination of the surface and cross sections and accelerated aging (SO_2 and UV) were carried out. The silicone-acrylic mixtures (Raccanello and Bologna Cocktail) gave better water repellency. However, the overall conclusion was that on these compact stones all of the treatments acted only as superficial protectives.

1983

Alessandrini, G., G. Dassu, and R. Peruzzi. Restoration of the facade of the Certosa di Garignano in Milan: Degradation and methods of conservation. Abstract only in *Fourth International Congress on the Deterioration and Preservation of Stone Objects, 7–9 July 1982,* ed. K. L. Gauri and J. A. Gwinn, 341. Louisville: University of Louisville, 1983. Mentions that Angera stone was consolidated with "silicon resins."

Amoroso, G. G., and V. Fassina. *Stone Decay and Conservation.* New York: Elsevier, 1983. Notes examples of silicone resins used for impregnating stone sculpture.

Andersson, T. Konservering av dopfunt i Fleringe kyrka, Gotland. In *Konserveringstekniska studier,* 140–44. Stockholm: Civiltryck AB, 1983. The stone of a baptismal font was consolidated with "silica ester."

de Castro, E. Studies on stone treatments. in *Fourth International Congress on the Deterioration and Preservation of Stone Objects, 7–9 July 1982*, ed. K. L. Gauri and J. A. Gwinn, 119–25. Louisville: University of Louisville, 1983. Studies the use of contact angle measurement and microdrop absorption on treated stone for determining the degree of water repellency of treatments before and after weathering. Compares acrylic and silicone treatments.

Cecchi, R. M., G. Alessandrini, G. Dassu, R. Peruzzi, G. Biscontin, and R. Cemesini. Restoration of the facade of the Certosi di Garignano in Milan. Abstract only in *Fourth International Congress on the Deterioration and Preservation of Stone Objects, 7–9 July 1982*, ed. K. L.Gauri and J. A. Gwinn, 341. Louisville: University of Louisville, 1983. Reports on the use of silicone resins for consolidation of the church facade.

Charola, A. E., G. Wheeler, and R. J. Koestler. Treatment of the Abydos reliefs: Preliminary investigations. In *Fourth International Congress on the Deterioration and Preservation of Stone Objects, 7–9 July 1982*, ed. K. L. Gauri and J. A. Gwinn, 77–88. Louisville: University of Louisville, 1983. Determines composition of stone and salt content and identifies previous treatments on this set of Egyptian limestone reliefs. Examines the effectiveness of several consolidation treatments including MTMOS.

Chiari, G. Characterization of adobe as building material: Preservation techniques. In *El adobe: Simposio internacional y curso-taller sobre conservación del adobe: Informe final y ponencias principales: Lima-Cusco (Peru), 10–22 September 1983*, 33–43. Lima: UNDP/UNESCO, 1983. Treatment of surfaces with ethyl silicate is mentioned.

Clifton, J. R., and M. Godette. Performance tests for stone consolidants. In *Fourth International Congress on the Deterioration and Preservation of Stone Objects, 7–9 July 1982*, ed. K. L.Gauri and J. A. Gwinn, 101–8. Louisville: University of Louisville, 1983. Review of 25 stone consolidants ranging from alkoxysilanes to epoxy resins by tensile strength, compressive strength, artificial weathering (acid immersion), and freeze-thaw testing.

Domaslowski, W. Consolidation of the deteriorated portions of stone in historical monuments. Abstract only in *Fourth International Congress on the Deterioration and Preservation of Stone Objects, 7–9 July 1982*, ed. K. L. Gauri and J. A. Gwinn, 341. Louisville: University of Louisville, 1983. Wacker OH applied to a fine-grained limestone is found to impart hydrophobic properties to the stone that decreased over time.

Fritisch, H., E. Schamberg, C. Ceccarelli, F. Franzetti, and E. Pozzi. Conservation of natural stone. *Pitture e Vernici* 59(3):39–43. [Italian] Reports on performance of silicone-derived products on sandstone.

Gale, F. R., and N. R. Weiss. A study of examination and treatment techniques for a limestone gazebo. In *Fourth International Congress on the Deterioration and Preservation of Stone Objects, 7–9 July 1982*, ed. K. L. Gauri and J. A. Gwinn, 135–45. Louisville: University of Louisville, 1983. Tested ACRYLOID B67 in petroleum ether, alkoxysilanes (Union Carbide, TEOS, and MTMOS), and barium hydroxide on limestone.

Gamarra, R. M. Conservation of structures and adobe decorative elements in Chan Chan. In *El adobe: Simposio internacional y curso-taller sobre conservación del adobe: Informe final y ponencias principales: Lima-Cusco (Peru), 10–22 September 1983*,

109–15. Lima: UNDP/UNESCO, 1983. Reports on the use of ethyl silicate for surface consolidation of adobe.

Jaton, C., A. Bouineau, and R. Coignard. Experimental treatment of rocks. In *Fourth International Congress on the Deterioration and Preservation of Stone Objects, 7–9 July 1982*, ed. K. L. Gauri and J. A. Gwinn, 205–18. Louisville: University of Louisville, 1983. [French] Limestone, sandstone and granite were treated with, alkoxysilane, epoxy, polyester, or polyacrylamide resin and evaluated for porosity and pore sizes.

Kotlik, P., and J. Zelinger. Physical properties of sandstone on impregnation with organosilicon consolidants and epoxy resin. *Sbornik Vysoke školy chemicko-technologichke v Praze* S10:129–44. [English with Czech and German summaries] "Organosilicon" and epoxy resin consolidants were applied to sandstone. The organosilicon penetrated better, changed color less, and had better vapor permeability. Neither treatment stops the movement of soluble salts in the pores.

Králová, H., and J. Šrámek. The protective ability of some materials used for the preservation of stone against the attack of sulfur dioxide. *Sborník restaurátorskych praci*, vol. 1:106–11. [Czech with English and German summaries] The effect of sulfur dioxide on a calcareous marly limestone was lowered by treatments of acrylates, polyesters, and siloxanes.

Larson, J. H. A museum approach to the techniques of stone conservation. In *Fourth International Congress on the Deterioration and Preservation of Stone Objects, 7–9 July 1982*, ed. K. L. Gauri and J. A. Gwinn, 219–37. Louisville: University of Louisville, 1983. Outlines the development of stone conservation treatments at the Victoria and Albert Museum. Outlines the use of MTMOS (Rhone-Poulenc X54-802 and Dow-Corning T-40149 with catalyst titanium iso-propoxide) TEOS (ICI EP 5850) Wacker and Tegovakon products (as well as several organic resins and mixtures) on marbles, English sandstones, and English limestones. Claims that the catalyzed systems were generally unstable and tended to swell and disrupt the stone during retreatment.

Laurenzi Tabasso, M. Trattamenti di conservazione sul marmo. In *Marmo restauro: situazione e prospettive, atti del convegno, Carrara, 31 maggio 1983*, 1983, 71–82. Review of marble conservation treatments including ethyl silicates. Consolidants are listed, and proper use is described.

Lindenthal, F. Die Apostelfiguren am Petersportal des Kölner Domes. *Jahrbuch der Rheinischen Denkmalpflege* 29:281–86. After cleaning, limestone statues were repeatedly treated with ethyl silicate until completely saturated.

Muzovskaya, O. A., I. B. Timofeeva, and M. M. Averkina. Use of organosilicon compounds in restoration. *Novye oblasti primeneniya metallorganicheskikh soedinenii*, 1983, 168–76. [Russian] Discusses use of organosilicon-based products for strengthening stone.

Snethlage, R. Die Wasserdampf-Sorptionseigenschaften von Standsteinen und ihre Bedeutung für die Konservierung. In *Werkstoffwissenschaften und Bausanierung: Berichtsband des internationalen Kolloquiums, Esslingen, 6–8 Sept. 1983*, ed. F. H. Wittman, 297–303. Lack & Chemie, 1983. Study of alkoxysilane-derived gels forming in contact with clay minerals.

Snethlage, R. Zeitraffende Labortestreihen an ausgewaehlten Sandsteinen. *Arbeitsblaetter für Restauratoren* 16(2), Group 6:168–76. Sandstone was treated with ethyl silicate consolidants and water-repellent products. Samples were then subjected to salt crystallization and dilute acids. The most effective treatment was the combination of a consolidant and a water repellent.

Szabo, E. F., and G. T. Szabo Jr. Cleaning and restoration of the California Building, Balboa Park, San Diego, California. In *Fourth International Congress on the Deterioration and Preservation of Stone Objects, 7–9 July 1982,* ed. K. L. Gauri and J. A. Gwinn, 321–33. Louisville: University of Louisville, 1983. Describes conservation treatment.

Wittman, F. H., and R. Prim. Mesures de l'effet consolidant d'un produit de traitement. *Matériaux et Construction* 16(94):235–42. Describes new method of assessing consolidant by measuring biaxial strengths of circular disks. Tests are based on evaluation of "silicic ester" treatments on stone.

Zador, M. New experiments on conservation of stone in Hungary. Abstract only in *Fourth International Congress on the Deterioration and Preservation of Stone Objects, 7–9 July 1982,* ed. K. L. Gauri and J. A. Gwinn, 344. Louisville: University of Louisville, 1983. Reviews surface and consolidation treatments.

1984

Ageeva, E. N., and E. A. Semenova. Use of organosilicon compounds for preserving marble sculpture in the open air. In *Kremniiorganicheskie soedineniia i materialy na ikh osnove: Trudy V soveshchaniia po khimii i prakticheskomu primeneniiu kremniiorganicheskikh soedinenii, Leningrad, 16–18 dekabria 1981,* ed. V. O. Reikhsfel'd, 249–52. Nauka, Leningradskoe otd-nie, 1984. [Russian] A number of organosilicon compounds were examined for the conservation of marble.

Alessandrini, G., G. Dassu, R. Bugnini, and L. Formica. The technical examination and conservation of the portal of St. Aquilino's chapel in the basilica of St. Lorenzo, Milan. *Studies in Conservation* 29(4):161–71. The Carrara marble portal was consolidated with a methylphenylsilicone resin.

Anon. Manufacturing agreement for Brethane. *BRE News,* no. 61(winter). Reports on the agreement between the Building Research Establishment and Colebrand Ltd. for the manufacture of Brethane.

Charola, A. E., R. Rossi-Manaresi, R. J. Koestler, G. Wheeler, and A. Tucci. SEM examination of limestones treated with silanes or prepolymerized silicone resin in solution. In *Adhesives and Consolidants,* ed. N. S. Brommelle, E. M. Pye, P. Smith, and G. Thomson, 182–84. London: IIC, 1984. Limestones were treated with MTMOS, Dri Film 104, ACRYLOID B72/MTMOS, and ACRYLOID B72/Dri Film 104. MTMOS reportedly bonded to stone, forming a network around crystals, and Dri Film 104 was reversible after several years.

Charola, A. E., G. Wheeler, and G. G. Freund. The influence of relative humidity in the polymerization of methyltrimethoxysilane. In *Adhesives and Consolidants,* ed. N. S. Brommelle, E. M. Pye, P. Smith, and G. Thomson, 177–81. London: IIC, 1984. The polymerization of neat MTMOS was found to be dependent on the relative humidity. Higher RH had shorter gel times and the gel more cracks; lower RH had a longer gel time, was lower in polymer yield, and had less cracking. MTMOS consolidated fine quartz well, coarse quartz less well, and marble powder not at all.

Clifton, J. R. Laboratory evaluation of stone consolidants. In *Adhesives and Consolidants,* ed. N. S. Brommelle, E. M. Pye, P. Smith, and G. Thomson, 151–55. London: IIC, 1984. Treated limestone and sandstone with a number of consolidants, including alkoxysilanes.

Fox, J. C. Choosing a consolidant for limestone: Preliminary consideration. In *Papers Presented at the Art Conservation Training Programs Conference, 55–74.* Cambridge, Mass.: Harvard University Art Museums, 1984. Reviews criteria for the consolidation of salt-laden limestone.

Hanna, S. B. The use of organo-silanes for the treatment of limestone in an advanced state of deterioration. In *Adhesives and Consolidants,* ed. N. S. Brommelle, E. M. Pye, P. Smith, and G. Thomson, 171–76. London: IIC, 1984. Reviews alkoxysilane systems used at the British Museum: Dow-Corning T-40149 (MTMOS) + Raccanello E55050 (uncatalyzed acrylic-silane) and Wacker OH.

Larson, J., and J. Dinsmore. The treatment of polychrome medieval English stone sculpture in the museum environment. In *Adhesives and Consolidants,* ed. N. S. Brommelle, E. M. Pye, P. Smith, and G. Thomson, 167–70. London: IIC, 1984. Discusses the use of acrylic-alkoxysilane for deep consolidation of stone sculptures in the Victoria and Albert Museum.

Nikitin, M. K., and S. A. Shadrin. Use of organosilicon materials in restoration of monuments to history and culture. In *Kremniiorganicheskie soedineniia i materialy na ikh osnove: Trudy V soveshchaniia po khimii i prakticheskomu primeneniiu kremniiorganicheskikh soedinenii, Leningrad, 16–18 dekabria 1981,* ed. V. O. Reikhsfel'd, 231–38. Nauka, Leningradskoe otd-nie, 1984. [Russian] Lists organosilicones used in stone conservation.

Nishiura, T., M. Fukuda, and S. Miura. Treatment of stone with synthetic resins for its protection against damage by freeze-thaw cycles. In *Adhesives and Consolidants,* ed. N. S. Brommelle, E. M. Pye, P. Smith, and G. Thomson, 156–59. London: IIC, 1984. Compared a prepolymerized MTEOS (Colcoat SS-101) to ACRYLOID B72 and Araldite epoxy resin consolidants on volcanic tuff. The alkoxysilane held up much better than the other treatments after five cycles of freeze-thaw testing and gave better hydrophobicity. The epoxy resin gave the best strength increases.

Perander, T., and T. Raman. Deterioration of old brick work due to salt effects and brickwork preservation. In *International Conference on Durability of Building Materials and Components, Espoo, Finland, August 12–15, 1984,* vol. 2, no. 3.2, 1984, 463–77. Discusses alkoxysilanes treatments on salt-laden brick in laboratory and field trials.

Rusges, W. Stone cleaning and preservation. *Deutsche Malerblatt* 55(12):1354–56. [German] Reviews cleaning treatments and consolidants for stone.

Weber, H. Conservation and restoration of natural stone. *Kunststoffe im Bau* 19(2):67–71. [German] Discusses the use of ethyl silicate stone treatments.

Wheeler, G., J. Dinsmore, L. J. Ransick, A. E. Charola, and R. J. Koestler. Treatment of the Abydos reliefs: Consolidation and cleaning. *Studies in Conservation* 29(1):42–48. Discusses the use of alkoxysilanes as preconsolidants and consolidants for Egyptian limestone.

1985

Alessandrini, G., R. Bugini, E. Broglia, G. Dassu, and R. Peruzzi. 1.) S. Maria delle Grazie: Chiostro delle rane (Milano); 2.) Castello di Vigevano: Falconiera; 3.) Chiesa di S. Agostino a Bergamo. *Arte Lombarda,* no. 72:1–12. Sound limestone was treated with a silicone consolidant and a silicone water repellent, and decayed sandstone was treated with ethyl silicate and silicone water repellent.

Bradley, S. M. Evaluation of organo silanes for use in consolidation of sculpture displayed indoors. In *Fifth International Congress on the Deterioration and Conservation of Stone, Proceedings, Lausanne, 25–27 September 1985*, ed. G. Felix and V. Furlan, 759–67. Lausanne: Presses Polytechniques Romandes, 1985. Investigation of alkoxysilane stone consolidants on deteriorated limestone. Consolidants tested were MTMOS, catalyzed MTMOS, MTMOS + Raccanello (acrylic-silane mixture), Wacker OH, Wacker H, and Raccanello. The effectiveness of the consolidant's performance was tested by immersing the samples in 20% HCl for 48 hours.

De Witte, E., A. E. Charola, and R. P. Sherryl. Preliminary tests on commercial stone consolidants. In *Fifth International Congress on the Deterioration and Conservation of Stone, Proceedings, Lausanne, 25–27 September 1985*, ed. G. Felix and V. Furlan, 709–18. Lausanne: Presses Polytechniques Romandes, 1985. Wacker H and OH and Tegovakon V and T were evaluated. All treated samples were hydrophobic three weeks after application, and Wacker OH and Tegovakon T lost this effect after one month. The films were etched and studied by SEM. The gels appeared as discrete spongy masses in the limestone.

Domaslowski, W., and E. Derkowska. Investigation on consolidation and stabilization of waterlogged argillaceous ground by means of electrokinetic effect. In *Fifth International Congress on the Deterioration and Conservation of Stone, Proceedings, Lausanne, 25–27 September 1985*, ed. G. Felix and V. Furlan, 719–26. Lausanne: Presses Polytechniques Romandes, 1985. Attempt to achieve better penetration of chemical treatments by use of electrokinetic techniques.

Koblischek, P. J. Polymers in the renovation of buildings constructed from natural stone. *Kunstoffe im Bau* 20(4):185–89. General discussion of stone consolidants, including alkoxysilanes.

Koestler, R. J., A. E. Charola, and G. Wheeler. Scanning electron microscopy in conservation: The Abydos reliefs. In *Application of Science in the Examination of Works of Art*, ed. P. A. England and L. van Zelst, 225–29. Boston: Museum of Fine Arts Research Laboratory, 1985. Decayed limestone that had been previously consolidated with paraffin and tung oil was treated with MTMOS and examined by SEM.

Kotlík, P., A. Faust-Lásló, and J. Zelinger. Silanol primers as applied in the preparation of artificial sandstone. *Sborník Vysoké školy chemicko-technologocé v Praze* 13:233–44. [Czech with English and German summaries] Artificial stone made from sand, "silanol" primers, and epoxy resin.

Kotlík, P., and J. Zelinger. Natural aging of the surface protection layers on the limestone. *Pamiatky príroda* 25(1):20–25. [Slovak] Waxes, Wacker H, and an epoxy resin were tested on a limestone. Natural weathering showed that Renaissance Wax, ACRYLOID B72, and Wacker H performed best.

Laurenzi Tabasso, M., and U. Santamaria. Consolidant and protective effects of different products on Lecce limestone. In *Fifth International Congress on the Deterioration and Conservation of Stone, Proceedings, Lausanne, 25–27 September 1985*, ed. G. Felix and V. Furlan, 697–707. Lausanne: Presses Polytechniques Romandes, 1985. Consolidated samples are artificially weathered under mild conditions. Wacker OH slightly decreases water absorption and gives good increases in compressive strength.

Lazzarini, L. *Some Experiences with the Conservation of Natural Stone in Venice*. Arbeitsheft no. 31. Munich: Bayerisches Landesamt für Denkmalpflege, 1985. Discusses treatments, including alkoxysiloxanes, used for stone conservation.

Lewin, S. Z., and G. Wheeler. Alkoxysilane chemistry and stone conservation. In *Fifth International Congress on the Deterioration and Conservation of Stone, Proceedings, Lausanne, 25–27 September 1985,* ed. G. Felix and V. Furlan, 831–44. Lausanne: Presses Polytechniques Romandes, 1985. Examines reactions of TEOS in water/ethanol solutions. Molar ratio of water to TEOS must be at least 2:1 to lead to gelation.

Mavrov, G. Technical investigation on the conservation of an eighth-century rock relief. In *Application of Science in the Examination of Works of Art,* ed. P. A. England and L. van Zelst, 230–33. Boston: Museum of Fine Arts Research Laboratory, 1985. Tested Drisil 773, MTMOS (Rhone-Poulenc X54-802), and Wacker OH on a Bulgarian rock relief.

NATO-CCMS. Characterization of stone types: Comparative assessment of water-repelling and stone-reinforcing agents. *Rijksdienst voor de Monumentenzorg,* Holland, 1985. Compilation of data on testing consolidants.

Nishiura, T. Conservation treatment of roof stone of Sunuiyamutaki stone gate. In *Conservation and Restoration of Stone Monuments: Scientific and Technical Study on the Conservation and Restoration of Monuments Made of Stone or Related Materials,* ed. Y. Emoto, T. Nishiura, and S. Miura, 134–39. Tokyo: Tokyo National Research Institute of Cultural Properties, 1985 [Japanese with English summary] Stones were treated with methyltriethoxysilane-based consolidant (Colcoat SS-101).

Nishiura, T. An experimental study on the mixture of silane and organic resin as consolidant of granularly decayed stone: Studies on the conservation treatment of stone (V). In *Conservation and Restoration of Stone Monuments: Scientific and Technical Study on the Conservation and Restoration of Monuments Made of Stone or Related Materials,* ed. Y. Emoto, T. Nishiura, and S. Miura, 105–10. Tokyo: Tokyo National Research Institute of Cultural Properties, 1985. [Japanese with English summary] Alkoxysilanes do not consolidate granularly disintegrated stone.

Nishiura, T. An experimental study on the treatment of old roof tiles for reuse by impregnation with silane. In *Conservation and Restoration of Stone Monuments: Scientific and Technical Study on the Conservation and Restoration of Monuments Made of Stone or Related Materials,* ed. Y. Emoto, T. Nishiura, and S. Miura, 111–19. Tokyo: Tokyo National Research Institute of Cultural Properties, 1985. [Japanese with English summary] Old roof tiles treated by immersion in a methyltriethoxysilane-based consolidant (Colcoat SS-101).

Nishiura, T. A practical study on the conservation treatment of old roof tiles of a historic building. In *Conservation and Restoration of Stone Monuments: Scientific and Technical Study on the Conservation and Restoration of Monuments Made of Stone or Related Materials,* ed. Y. Emoto, T. Nishiura, and S. Miura, 144–51. Tokyo: Tokyo National Research Institute of Cultural Properties, 1985. [Japanese with English summary] Old roof tiles treated by immersion in a methyltriethoxysilane-based consolidant (Colcoat SS-101).

Nishiura, T. The protection of rocks from frost damage with synthetic resins. In *Conservation and Restoration of Stone Monuments: Scientific and Technical Study on the Conservation and Restoration of Monuments Made of Stone or Related Materials,* ed. Y. Emoto, T. Nishiura, and S. Miura, 22–32. Tokyo: Tokyo National Research Institute of Cultural Properties, 1985. [Japanese with English summary] Oya stone is consolidated with epoxy resin, acrylic resin, and a methyltrimethoxysilane-based consolidant (Colcoat SS-101) and subjected to freeze-thaw cycles. This latter treatment alters the "hydraulic" conductivity of the rock and leads to damage during freezing.

Nishiura, T. Relationship between the moisture content of stone and the hydrophobic effect of silane treatment: Studies on the conservation treatment of stone (IV). In *Conservation and Restoration of Stone Monuments: Scientific and Technical Study on the Conservation and Restoration of Monuments Made of Stone or Related Materials,* ed. Y. Emoto, T. Nishiura, and S. Miura, 101–4. Tokyo: Tokyo National Research Institute of Cultural Properties, 1985. [Japanese with English summary] The moisture content of stone was found to affect the depth of penetration and the degree of hydrophobicity obtained from an oligomeric methyltriethoxysilane consolidant (Colcoat SS-101).

Nishiura, T. Salt crystallization decay of stone treated with resin-water evaporation from stone treated with silane and its salt crystallization decay: Studies on the conservation treatment of stone (II). In *Conservation and Restoration of Stone Monuments: Scientific and Technical Study on the Conservation and Restoration of Monuments Made of Stone or Related Materials,* ed. Y. Emoto, T. Nishiura, and S. Miura, 59–72. Tokyo: Tokyo National Research Institute of Cultural Properties, 1985. [Japanese with English summary] Hydrophobic alkoxysilane-derived consolidants impede water evaporation from the interior. Salts were found to crystallize under the hydrophobic layer and cause exfoliation.

Nishiura, T. Split test of stone treated with synthetic resins: Studies on the conservation treatment of stone (III). In *Conservation and Restoration of Stone Monuments: Scientific and Technical Study on the Conservation and Restoration of Monuments Made of Stone or Related Materials,* ed. Y. Emoto, T. Nishiura, and S. Miura, 97–100. Tokyo: Tokyo National Research Institute of Cultural Properties, 1985. [Japanese with English summary] Consolidated sound tuff with a number of products, including alkoxysilanes. Results were inconclusive, as no great increases in strength were found.

Pancella, R., and V. Furlan. Proprietés d'un grès traite tendre avec des résines synthetiques, 1ere, 2e parties. *Chantiers* 16(4–5) 295–98; 409–13. Evaluates the improvement in mechanical properties of calcareous sandstone by consolidation with a number of different products. Two ethyl silicate consolidants and an alkylmethoxysilane were among the consolidants tested.

Prim, P., and F. Wittmann. Méthode de mesure de l'effet consolidant de produits de traitement de la pierre. In *Fifth International Congress on the Deterioration and Conservation of Stone, Proceedings, Lausanne, 25–27 September 1985,* ed. G. Felix and V. Furlan, 787–95. Lausanne: Presses Polytechniques Romandes, 1985. Describes the method of biaxial stress testing for consolidated and unconsolidated stone discs.

Rosenfeld, A. *Rock Art Conservation in Australia.* Special Australian Heritage, no. 2. Canberra: Australian Government Publication Series, 1985. Reviews rock art conservation and mentions alkoxysilanes as consolidants.

Rossi-Manaresi, R. SEM examination of a biocalcarenite treated with acrylic polymers, silane, or silicone resins. In *Fifth International Congress on the Deterioration and Conservation of Stone, Proceedings, Lausanne, 25–27 September 1985,* ed. G. Felix and V. Furlan, 871–80. Lausanne: Presses Polytechniques Romandes, 1985. Studied consolidated samples by SEM. MTMOS had good penetration, but its films were too thin to provide good consolidation.

Snethlage, R. *Steinkonservierung—zum Stand der Forschung.* Arbeitsheft, no. 31. Munich: Bayerisches Landesamt für Denkmalpflege, 1985. Reviews of current consolidation practices.

Somlo, G. Protection of monuments with silicic acid ester. *Természet Világa* 116(2):559–60. [Hungarian] Weathering of stone is decreased by either spraying or consolidating with "silicic acid ester."

Weber, H. The conservation of the Alte Pinakothek in Munich. Laurenzi Tabasso, M., and U. Santamaria. Consolidant and protective effects of different products on Lecce limestone. In *Fifth International Congress on the Deterioration and Conservation of Stone, Proceedings, Lausanne, 25–27 September 1985*, ed. G. Felix and V. Furlan, 1063–72. Lausanne: Presses Polytechniques Romandes, 1985. Reviews testing program for calcareous sandstone and brick building.

Weber, H. *Steinkonservierung, Leitfaden zur Konservierung und Restaurierung von Natursteinen.* 3d ed. Wurttemburg: W. Bartz, 1985. Review of the conservation and preservation of stone.

Zinsmeister, K. J. H., N. R. Weiss, and F. R. Gale. Evaluation of consolidation treatment of an American sandstone. *Bautenschutz und Bausanierung* 8(2):79–82. [German] Treatment with ethyl silicate consolidants resulted in dramatic increases in compressive strength, modulus of rupture, and abrasion resistance of the sandstone samples.

1986

Alessandrini, G. San Michele in Pavia: Prove di laboratorio per la valutazione dell'efficacia di interventi conservative. In *La Pietra del San Michele: Restauro e conservazione*, 69–76. Pavia: Societa Conservazione Arte Cristiana, 1986. A number of products, including an ethyl silicate consolidant, were applied to stone samples. Tests included water absorption, weathering cycles, UV radiation, etc. A silicone resin gave the best results.

Bradley, S. M. An introduction to the use of silanes in stone conservation. *Geological Curator* 4(7):427–32. General review of chemistry, properties, and practical applications of alkoxysilane stone consolidants.

Bradley, S. M., and S. B. Hanna. The effect of soluble salt movement on the conservation of an Egyptian limestone standing figure. In *Case Studies in the Conservation of Stone and Wall Paintings*, ed. N. S. Brommelle and P. Smith, 57–61. London: IIC, 1986. Reports that soluble salts are not immobilized by alkoxysilane consolidants.

Charola, A. E., and R. J. Koestler. Scanning electron microscopy in the evaluation of consolidation treatments for stone. In *Scanning Electron Microscopy*, vol. 2, ed. R. P. Becker and G. M. Roomans, 479–84. Chicago: Scanning Electron Microscopy, Inc., 1986. Compared the Bologna Cocktail to gels formed from Wacker H and OH in limestone. The Wacker OH appeared as "sponge-like clumps" deposited between the grains.

Charola, A. E., L. Lazzarini, G. Wheeler, and R. J. Koestler. The Spanish apse from San Martin de Fuentidueña at the Cloisters, Metropolitan Museum of Art, New York. In *Case Studies in the Conservation of Stone and Wall Paintings*, ed. N. S. Brommelle and P. Smith, 18–21. London: IIC, 1986. Suggests ethyl silicate consolidants for the dolomitic limestone of the apse.

Charola, A. E., A. Tucci, and R. J. Koestler. On the reversibility of treatments with acrylic/silicone resin mixtures. *Journal of the American Institute for Conservation* 25(2):83–92. Samples of limestone treated with the Bologna Cocktail were artificially weathered in a sulfuric acid fog and cycling relative humidity and the degree of reversibility of the resins determined by immersion in solvent or by application of poultices. A comparison between the weathered and unweathered samples shows a decrease in the solubility.

Fritsch, H. Ripristino, mediante composti organici del siliciom della solidita originale della pietra naturale danneggiata da agenti atmosferici. In *Manutenzione e conservazione del costruito fra tradizione ed innovazione: Atti del convegno di studi, Bressanone, 24–27 giugno 1986*, ed. G. Biscontin, 513–25. Padova: Libreria Progetto Editore, 1986. Tested Tegovakon V and Tegosivin HL 100 on marlstone. Abrasion resistance measurements were compared to values of density and porosity of sound, deteriorated, and treated stone.

Králová, M., and J. Šrámek. Monitoring stability of some polymers used to treat stones. *Sborník restaurátorskych prací* 2–3:78–87. [Czech with Russian, English, and German summaries] Alkylalkoxysilanes were found to be the most stable product after being exposed to ultraviolet radiation and high sulfur dioxide concentration.

Larson, J. The conservation of a marble group of Neptune and Triton by Gian Lorenzo Bernini. In *Case Studies in the Conservation of Stone and Wall Paintings*, ed. N. S. Brommelle and P. Smith, 22–26. London: IIC, 1986. Describes consolidant testing program, and the rationale for using an acrylic-alkoxysilane mixture is explained. Treatment reevaluated after six years and found to be satisfactory.

Lazzarini, L., and M. Laurenzi Tabasso. *Il restauro della pietra.* Padova: CEDAM, 1986. Overview of stone conservation.

Lukaszewicz, J. Study of the properties of porous materials fortified with Steinfestiger OH preparation. *Rocznik przedsiebior stwa pantswo wego Pracownie Konserwacji zabytkow*, no. 1:81–99. [Polish with Russian and English summaries] Emphasizes influence of porous material type and conditions on the derived gel.

Mangum, B. On the choice of preconsolidant in the treatment of an Egyptian polychrome triad. In *Case Studies in the Conservation of Stone and Wall Paintings*, ed. N. S. Brommelle and P. Smith, 148–50. London: IIC, 1986. Use of alkoxysilanes on a polychromed sandstone causes darkening.

Nishiura, T. Study on the color change induced by the impregnation of silane. I. Analysis of the color change by color meter. Study on the conservation treatment of stone (VII). *Kobunkazai no kagaku (Scientific Papers on Japanese Antiques and Art Crafts)* 31(December):41–50. [Japanese with Japanese and English summaries] MTMOS changed the color of the treated stone by lowering the value. Changes to hue and chroma were negligible. After aging, the surfaces were almost the same color as the untreated. Japanese-language version of paper presented at 1987 ICOM meeting.

Torraca, G. Momenti nella storia della conservazione del marmo: Metodi e attitudini in varie epoche. In *OPD restauro: Quaderni dell'opificio dell pietra dure e laboratori di restauro di Firenze*, numero speciale, 1986, 32–45. [Italian] Reviews marble conservation and consolidants.

Torraca, G., and J. Weber. *Poröse Baustoffe: Eine Materialkunde für die Denkmalpflege.* Vienna: Verlag Der Apfel, 1986. Revised German edition of G. Torraca, *Porous Building Materials—Materials Science for Architectural Conservation* (1981).

1987

Andersson, T., P. Elfving, and M. Lerjefors. *Kiselsyraesterbaserad stenkonservering.* Götenborg: Chalmers Tekniska Hogskola, 1987. [Swedish] Examines the chemical bonding between stones and alkoxysilane-based consolidants and hydrophobing agents.

Biscontin, G., C. Botteghi, C. Vecchia, C. Dalla, G. Driussi, G. Moretti, and A. Valle. Stability study of siliconic resins employed in the stone conservation. In *ICOM Committee for Conservation, Eighth Triennial Meeting, Sydney, Australia, 6–11 September 1987: Preprints*, ed. K. Grimstad, 785–90. Los Angeles: Getty Conservation Institute, 1987. Examines Rhone-Poulenc products 11309 and XR893 by FT-IR. After aging, the OH region increases.

Biscontin, G., A. Masin, R. Angeletti, and G. Driussi. Influence of salts on the efficiency of a conservative siliconic treatment on stones. In *ICOM Committee for Conservation, Eighth Triennial Meeting, Sydney, Australia, 6–11 September 1987: Preprints*, ed. K. Grimstad, 469–73. Los Angeles: Getty Conservation Institute, 1987. Attempted surface and deep consolidation with methylphenylpolysiloxane on stone was saturated with NaCl and $MgSO_4$.

Capponi, G., and C. Meucci. Il restauro del paramento lapideo della facciata della chiesa di S. Croce a Lecce. In *Materiali lapidei: Problemi relativi allo studio delgrado e della conservazione*, ed. A. Bureca, M. Laurenzi Tabasso, and G. Palandri, 263–82. Rome: Istituto Poligrafico e Zecca dello Stato, 1987. Badly decayed porous limestone consolidated with ethyl silicate.

Chiari, G. Consolidation of adobe with ethyl silicate: Control of long-term effects using SEM. In *Fifth International Meeting of Experts on the Conservation of Earthen Architecture*, 25–32. Rome: ICCROM-CRATerre, 1987. SEM photographs showing long, weblike strands of ethyl silicate–derived gels in adobe.

Dinsmore, J. Considerations of adhesion in the use of silane-based consolidants. *The Conservator* 11:26–29. Reports on the use of adhesives with neat MTMOS on limestone.

Esbert, R. M., C. Grossi, and R. M. Marcos. Laboratory studies on the consolidation and protection of calcareous materials in the Cathedral of Oviedo. Part II. *Materiales de Construcción* 37(308):13–21. A dolomite treated with ethyl silicate and an oligomeric alkoxysilane performed better under freeze-thaw and salt crystallization testing than did the sample treated with a silicone resin.

Fassina, V., M. Laurenzi Tabasso, L. Lazzarini, and A. M. Mecchi. Protective treatment for the reliefs of the Arconi di San Marco in Venice: Laboratory evaluation. *Durability of Building Materials* 5(2):167–81. Acrylic-silicone mixtures were tried, along with other consolidants, on two marble types. All treatments improved the hydrophobicity but reduced the vapor permeability.

Fazio, G. Sull'efficacia di alcuni trattamenti di restauro realizzati dopo il 1960. In *Materiali lapidei: Problemi relativi allo studio delgrado e della conservazione*, ed. A. Bureca, M. Laurenzi Tabasso, and G. Palandri, 197–214. Rome: Istituto Poligrafico e Zecca dello Stato, 1987. Compares acrylic, silicone, and epoxy resin consolidants.

Ganorkar, M. C., T. A. Sreenivasa Rao, and M. Bhaskar Reddy. Deterioration and conservation of calcareous stones. In *ICOM Committee for Conservation, Eighth Triennial Meeting, Sydney, Australia, 6–11 September 1987: Preprints*, ed. K. Grimstad, 479–86. Los Angeles: Getty Conservation Institute, 1987. Suggests new application methods for inorganic consolidants. Also reports that inorganic polymers are effective in arresting deterioration.

Koestler, R. J., E. Santoro, F. Preusser, and A. Rodarte. A note on the reaction of methyl tri-methoxysilane to mixed cultures of microorganisms. In *Biodeterioration Research 1, Proceedings of the First Annual Meeting of the Pan American*

Biodeterioration Society, Held July 17–19, 1986, Washington, D.C., ed. G. C. Llewellyn and C. E. O'Rear, 317–21. New York: Plenum, 1987. MTMOS was exposed to mixed cultures of a fungus, a cyanobacterium, and an algae and examined by SEM and FTIR. Concluded that MTMOS would not encourage growth on objects in museums because, although SEM showed some biological growth, FTIR showed no evidence of any changes in bond structures.

Kotlík, P. Chemical problems of consolidation of stone with organosilicon consolidants. *Chemicke listy* 81(5):511–15. [Czech with English summary] Reviews alkoxysilane hydrolysis reactions. Highlights that alkaline components affect the condensation of silanol groups.

Molteni, C. Rhodorsil. *Pitture e Vernici* 63(8):13–33. General review of organosilicon products used in stone conservation.

Nishiura, T. Laboratory evaluation of the mixture of silane and organic resin as consolidant of granularly decayed stone. In *ICOM Committee for Conservation, Eighth Triennial Meeting, Sydney, Australia, 6–11 September 1987: Preprints,* ed. K. Grimstad, 805–7. Los Angeles: Getty Conservation Institute, 1987. Consolidated granulated tuff with (1) methyltrimethoxysilane-based consolidant (Colcoat SS-101); (2.) N-β-(aminoethyl)-aminopropylmethyldimethoxysilane; (3) γ-methacryloxypropyltrimethoxysilane. The alkoxysilanes were applied neat or as coupling agents for ACRYLOID B72 or an epoxy resin. The coupling agents improved the weathering ability of the organic resins.

Nishiura, T. Laboratory test on the color change of stone by impregnation with silane. In *ICOM Committee for Conservation, Eighth Triennial Meeting, Sydney, Australia, 6–11 September 1987: Preprints,* ed. K. Grimstad, 509–12. Los Angeles: Getty Conservation Institute, 1987. Applied oligomeric MTEOS (Colcoat SS-101) to porous tuff. The color change in the consolidated stone was proportionate to the concentration of the alkoxysilane in the treatment solution and was mainly a fall in value (darkening) rather than any change in chroma or hue. After accelerated weathering, the darkening diminished to approximately that of the untreated stone.

Plog, C., W. Kerfin, G. Roth, W. Gerhard, and H. Weber. Surface analysis studies using SIMS for preservation of natural rock. *Bautenschutz und Bausanierung* 10(3):127–30. SIMS is used to evaluate the penetration of consolidants into stone.

Robinson, H. L. Durability of anti-carbonation coatings. *Journal of the Oil and Colour Chemists' Association* 70(7):193–98. Tests a number of treatments as to their gas barrier properties.

Robinson, H. L. An evaluation of silane treated concrete. *Journal of the Oil and Colour Chemists' Association* 70(6):163–72. Monomeric alkoxysilane found to give the best resistance to the diffusion of chloride.

Roth, M. Die Wirksamkeit einer Siliconimprägnierung. *Bautenschutz und Bausanierung* 10(4):149–50. Silicones with longer alkyl groups proved more effective than methyl-silicone. Methylsilicone degraded after five years on limestone.

Tubb, K. W. Conservation of the lime plaster statues of 'Ain Ghazal. In *Recent Advances in the Conservation and Analysis of Artifacts: Jubilee Conservation Conference, London 6–10 July 1987,* 387–92. London: Summer Schools Press, 1987. Discussion centers on the choice of an appropriate consolidant for the statues, pointing out the advantages and disadvantages of the Raccanello E55050 and Dow-Corning Z6070 (MTMOS).

Wheeler, G. The chemistry of four alkoxysilanes and their potential for use as stone consolidants. Ph.D. dissertation, New York University, 1987. Photocopy. Ann Arbor, Mich.: University Microfilms International. Examines the polymerization of aqueous solutions of TEOS, MTEOS, TMOS, and MTMOS by GC-MS, HPLC, and IR spectroscopy.

1988

Alessandrini, G., R. Bonecchi, E. Broglia, R. Bugini, R. Negrotti, R. Peruzzi, L. Toniolo, and L. Formica. Les colonnes de San Lorenzo (Milan, Italie): Identification des matériaux, causes d'altération, conservation. In *The Engineering Geology of Ancient Works, Monuments and Historic Sites: Proceedings of an International Symposium Organized by the Greek National Group of IAEG, Athens, 19–23 September 1988*, vol. 2, ed. P. Marinos and G. Koukis, 925–32. Rotterdam: A. A. Balkema, 1988. Reviews tests of different treatments on marble.

Ausset, P., and J. Philippon. Essai d'évaluation de profondeur de pénétration de consolidants de la pierre. In *Sixth International Congress on Deterioration and Conservation of Stone: Proceedings, Torun, 12–14 September 1988*, ed. J. Ciabach, 524–33. Torun: Nicholas Copernicus University, 1988. SEM used to determine distribution of silicon-based treatments in calcareous rocks.

Bell, F. G., and M. Coulthard. Stone preservation with illustrative examples from the United Kingdom. In *The Engineering Geology of Ancient Works, Monuments and Historic Sites: Proceedings of an International Symposium Organized by the Greek National Group of IAEG, Athens, 19–23 September 1988*, vol. 2, ed. P. Marinos and G. Koukis, 883–89. Rotterdam: A. A. Balkema, 1988. Review of stone preservation treatments with illustrative examples from the United Kingdom.

Boulton, A. Some considerations in the treatment of archaeological plaster figures from 'Ain Ghazal, Jordan. In *Preprints of Papers Presented at the 16th Annual Meeting of the American Institute for Conservation, New Orleans, June 1988*, ed. S. Z. Rosenberg, 38–57. Washington, D.C.: AIC, 1988. Reports on the mechanical testing program on plaster consolidated with ACRYLOID B72 and alkoxysilanes.

Charola, A. E. Brief introduction to silanes, siloxanes, silicones, and silicate esters. In *The Deterioration and Conservation of Stone: Notes from the International Venetian Courses on Stone Restoration*, ed. L. Lazzarini and R. Pieper, 313–16. Venice: UNESCO, 1988. Reviews basic chemistry of alkoxysilanes and their use on stone.

De Witte, E., A. Terfve, R. J. Koestler, and A. E. Charola. Conservation of the Goreme rock: Preliminary investigations. In *Sixth International Congress on Deterioration and Conservation of Stone: Proceedings, Torun, 12–14 September 1988*, ed. J. Ciabach, 346–55. Torun: Nicholas Copernicus University, 1988. Reports on artificial aging of treated samples of tuff.

Domaslowski, W., and J. W. Lukaszewicz. Possibilities of silica application in consolidation of stone monuments. In *Sixth International Congress on Deterioration and Conservation of Stone: Proceedings, Torun, 12–14 September 1988*, ed. J. Ciabach, 563–76. Torun: Nicholas Copernicus University, 1988. Steinfestiger OH applied to a limestone and sandstone. The bending strengths increased with a rise in the relative humidity during application. The gel did not form a coating but clusters of cracked gel in the pores.

Esbert, R. M., C. Grossi, L. M. Suarez del Rio, L. Calleja, J. Ordaz, and M. Montoto. Acoustic emission generated in treated stones during loading. In *Sixth International Congress on Deterioration and Conservation of Stone: Proceedings, Torun, 12–14 September 1988*, ed. J. Ciabach, 403–10. Torun: Nicholas Copernicus University,

1988. Evaluated treatment of ethyl silicate and oligomeric polysiloxanes on dolomitic stone by application of uniaxial compressive stresses in order to monitor acoustic emissions.

Heidingsfeld, V., M. Brabec, and J. Zelinger. Some characteristics of silicon oxide colloidal dispersion from the point of view of its application in plaster consolidation. *Sborník Vysoké školy chemicko-technologické v Praze, Polymery-chemie, vlastnosti a zpracování* S18:129–36. [Czech] Properties of Tosil (colloidal silicon oxide) are discussed. Gel times are found to depend on pH and be significantly affected by soluble salts.

Heidingsfeld, V., M. Brabec, and J. Zelinger. Study of the possibilities of applying ethylsilicate 40 as a plaster consolidant. *Sborník Vysoké školy chemicko-technologicke v Praze, Polymery-chemie, vlastnosti a zpracovaní* S18:117–28. [Czech] Studied gel quality, time to gelation, and effect of catalyst type on ethyl silicate 40 consolidant penetration.

D. Honsinger, and H. R. Sasse. New approaches of preservation of sandstone surfaces by polymers. *Bautenschutz und Bausanierung* 11(6):205–11. [German] General review of approaches to consolidation.

Koestler, R. J., and E. D. Santoro. *Assessment of the Susceptibility to Biodeterioration of Selected Polymers and Resins: Technical Report for the Getty Conservation Institute*. Marina del Rey, Calif: Getty Conservation Institute, 1988. Assesses microbial susceptibility of coatings and ranked them based on sensitivity to fungal deterioration.

Koestler, R. J., E. Santoro, J. Druzik, F. Preusser, L. Koepp, and M. Derrick. Status report: Ongoing studies of the susceptibility of stone consolidants to microbiologically induced deterioration. In *Biodeterioration 7: Selected Papers Presented at the Seventh International Biodeterioration Symposium, Cambridge, 6–11 September 1987*, ed. D. R. Houghton, R. N. Smith, and H.O.W. Eggins, 441–48. London: Elsevier, 1988. Studied stone consolidant films as to their susceptibility to biologically induced decay.

Kotlík, P., M. DvoWák, and J. Zelinger. Characteristics of gels SiO$_2$-polymers. *Sborník Vysoké školy chemicko-technologicke v Praze, Polymery-chemie, vlastnosti a zpracovaní* S18:177–85. Studies combination ethyl silicate 40/epoxy resin and ethyl silicate 40/ACRYLOID B72–derived gels. Ethyl silicate 40/B72 gives good results. Quality of gel is affected by catalyst type.

Laurenzi Tabasso, M. Conservation treatments of stone. In *The Deterioration and Conservation of Stone: Notes from the International Venetian Courses on Stone Restoration*, ed. L. Lazzarini and R. Pieper, 271–90. Venice: UNESCO, 1988. Reviews alkoxysilane treatments on stone.

Laurenzi Tabasso, M., A. M. Mecchi, U. Santamaria, and G. Venturi. Consolidation of stone 1: Comparison between a treatment with a methacrylic monomer and the corresponding polymer. In *The Engineering Geology of Ancient Works, Monuments and Historic Sites: Proceedings of an International Symposium Organized by the Greek National Group of IAEG, Athens, 19–23 September 1988*, vol. 2, ed. P. Marinos and G. Koukis, 933–38. Rotterdam: A. A. Balkema, 1988. Evaluation of methylmethacrylate and alkoxysilane consolidants on Carrara marble and volcanic tuff.

Laurenzi Tabasso, M., A. M. Mecchi, U. Santamaria, and G. Venturi. Consolidation of stone 2: Comparison between treatments with a methacrylic monomer and with alkoxysilanes. In *The Engineering Geology of Ancient Works, Monuments and Historic Sites: Proceedings of an International Symposium Organized by the Greek National Group of IAEG, Athens, 19–23 September 1988,* vol. 2, ed. P. Marinos and G. Koukis, 939–43. Rotterdam: A. A. Balkema, 1988. Compared treatments by measuring different physical parameters on Carrara marble and volcanic tuff.

Lewin, S. Z. The current state of the art in the use of synthetic materials for stone conservation. In *The Deterioration and Conservation of Stone: Notes from the International Venetian Courses on Stone Restoration,* ed. L. Lazzarini and R. Pieper, 290–302. Venice: UNESCO, 1988. Reviews barium hydroxide and "silicon ester" techniques.

Marchesini, L., and R. Bonora. The use of silicone resin in the conservation and protection of stone. In *The Deterioration and Conservation of Stone: Notes from the International Venetian Courses on Stone Restoration,* ed. L. Lazzarini and R. Pieper, 317–28. Venice: UNESCO, 1988. Reviews some alkoxysilane treatments on stone.

Riederer, J. The decay and conservation of marble on archaeological monuments. In *NATO ASI, Series E. Applied Sciences,* vol. 153, 465–74. Boston: M. Nijhoff, 1988. Discusses the use of acrylic resins versus ethyl silicate. Long-term evaluation was recommended to determine best performance.

Roth, M. Possibilities to render adobe water repellent. *Bautenschutz und Bausanierung* 11(2):76–77. [German] Methyltriethoxysilane in white spirits was found to be one of the best hydrophobic agents of mud-brick, along with potassium methylsiliconate and diluted oligomeric polysiloxanes.

Sattler, L., and R. Snethlage. Durability of stone consolidation treatments with silicic acid ester. In *The Engineering Geology of Ancient Works, Monuments and Historic Sites: Proceedings of an International Symposium Organized by the Greek National Group of IAEG, Athens, 19–23 September 1988,* vol. 2, ed. P. Marinos and G. Koukis, 953–56. Rotterdam: A.A. Balkema, 1988. Determined biaxial flexural strength of drill core slices. Cores were removed from consolidated objects and compared to samples treated in the laboratory. The cores removed from objects were found to have had low penetration, but the strength of the zones that were consolidated had not decreased in mechanical strength in nine years.

Stepien, P. Case studies in the use of organo-silanes as consolidants in conjunction with traditional lime technology. In *Sixth International Congress on Deterioration and Conservation of Stone: Proceedings, Torun, 12–14 September 1988,* ed. J. Ciabach, 647–52. Torun: Nicholas Copernicus University, 1988. Porous limestone and calcitic sandstone were first treated with limewater (limestone) and Wacker OH (sandstone) followed by cleaning, salt removal, and further treatment with Wacker OH.

Torraca, G. General philosophy of stone conservation. In *The Deterioration and Conservation of Stone: Notes from the International Venetian Courses on Stone Restoration,* ed. L. Lazzarini and R. Pieper, 243–70. Venice: UNESCO, 1988. General review of conservation techniques.

Weber, H., and H. Hohl. Verfahren zur Bestimmung der Eindringtiefe von Steinfestigungsmitteln auf der Basis von Kieselsaureester-Verbindungen. *Bautenschutz und Bausanierung* 11(6):200–204. [German] Uses staining technique to determine the depth of penetration of Wacker OH and H products.

Wheeler, G. The use of GC-MS in the study of alkoxysilane stone consolidants. In *Sixth International Congress on Deterioration and Conservation of Stone: Proceedings, Torun, 12–14 September 1988*, ed. J. Ciabach, 607–13. Torun: Nicholas Copernicus University, 1988. Used GC-MS to study reactions of TEOS and TMOS in alcohol-water solutions. TEOS has a much slower hydrolysis in neutral solutions than TMOS. To be of practical use, TEOS must be catalyzed or the water content increased to at least a ratio of 2:1. The co-solvent is also important. Reaction of TMOS was slowed considerably when ethanol, rather than methanol, was used.

Zielecka, M., and P. Rosciszewski. Water-repellent and structurally strengthening agents for masonry materials. In *Sixth International Congress on Deterioration and Conservation of Stone: Proceedings, Torun, 12–14 September 1988*, ed. J. Ciabach, 641–44. Torun: Nicholas Copernicus University, 1988. Ahydrosil Z, a 10% solution of modified methylsilicone resin in white spirits, increased the bending strength of limestone by 60–90% and sandstone by 80%.

Zinsmeister, K. H. J., N. R. Weiss, and F. R. Gale. Laboratory evaluation of consolidation treatment of Massillon (Ohio) sandstone. *APT Bulletin* 20(3):35–39. English version similar to Zinsmeister et al. 1985.

1989

Ettl, H., and H. Schuh. Konservierende Festigung von sandsteinen mit kiesel-säureethylester. *Bautenschutz und Bausanierung* 12:35–38. Discusses the overconsolidation of a surface with ethyl silicate and the resulting problems.

Lazzarini, L., M. Laurenzi Tabasso, and J. Philippon. *La restauration de la pierre*. Maurecourt: ERG, 1989. French translation of *Il restauro della pietra*, an overview of stone conservation.

Lukaszewicz, J. The influence of aging of polyalkoxysilane gel on the properties of stone consolidated with silicon esters. In *Sixth International Congress on Deterioration and Conservation of Stone, Supplement*, ed. J. Ciabach, 182–93. Torun: Nicholas Copernicus University, 1989. Investigates limestone and sandstone treated with Wacker OH.

Muller, V., H. Winkler, and D. Schmalstieg. Die Sicherung des Turmhelmes der Ev.-luth. Christuskirche in Hannover. In *Restaurierung von Kulturdenkmalen: Beispiele aus der niedesachsischen Denkmalpflege*, ed. H. Moller, 97–114. Berichte zur Denkmalpflege in niedersächsen 2. Hameln: Niemeyer, 1989. [German] Impregnated brick and sandstone with ethyl silicate using a tube system.

Price, C. Conservation de la façade ouest de la cathédrale de Wells. In *L'Ornementation architecturale en pierres dans les monuments historiques: Château de Fontainebleau, October 1988. Actes des colloques de la Direction du Patrimoine*, 1989, 38–40. Reviews consolidation of limestone with alkoxysilanes and lime treatments.

Roth, M. Il profilo di assorbimento d'acqua di una impregnazione silionica. In *Il Cantiere della Conoscenza: Il Cantiere del restauro: Atti del convegno di studi, Bressanone, 27–30 guigno 1989*, ed. G. Biscontin, M. Dal Colle, and S. Volpin, 417–26. Padova: Libreria Progetto Editore, 1989. [Italian with English summary] Reports Wacker Chemie tests on consolidant distribution in calcareous sandstone. Water repellency was found to decrease from the surface to the interior.

Weiss, N. R. How to restore stone: Methods of cleaning, pointing, repairing, treating, patching, replacing, and sealing architectural stone. *CRM Bulletin* 12(2):15–17. General review of treatments. Alkoxysilanes noted in the "use with caution section."

Wolter, H. Eigenschaften und Anwendungsgebiete von siliciumorganischen verbindungen bei der Hydrophobierung und Verfestigung von mineralischen Baustoffen. *Bautenschutz und Bausanierung* 12(1):9–14. [German] Reviews organosilanes for waterproofing and consolidation.

1990

Alessandri, P. M. Anfiteatro Flavio: Interventi conservativi sugli stucchi dell'ingresso nord. In *Superfici dell'architettura: Le finiture. Atti del convegno di studi, Bressanonne, 26–29 giugno 1990,* ed. G. Biscontin and S. Volpin, 367–76. Padova: Libreria Progetto Editore, 1990. [Italian with English summary] Stuccos on the entrance to the Roman Coliseum preconsolidated with ethyl silicate.

Alessi, P., A. Cortesi, G. Torriano, and D. Visintin. Problems connected with protection of monuments from aggressive agents. In *Conservation of Monuments in the Mediterranean Basin,* ed. F. Zezza, 383–87. Brescia: Grafo Edizioni, 1990. Tested a number of protective treatments on marble. Alkoxysilane treatments gave lower water absorption.

Ashurst, J., and F. G. Dimes. *Conservation of Building and Decorative Stone.* Vol. 2. Boston: Butterworth-Heinemann, 1990. Recommends the application alkoxysilanes to dry, or at least surface-dry, stone.

Brinker, C. J., and G. Scherer. *Sol-Gel Science: The Physics and Chemistry of Sol-Gel Processing.* Boston: Academic Press, 1990. Comprehensive review text on sol-gel science.

Chiari, G. Chemical surface treatments and capping techniques of earthen structures: A long-term evaluation. In *Proceedings from the Sixth International Conference on the Conservation of Earthen Architecture: Adobe 90 Preprints, Las Cruces, New Mexico, U.S.A., 14–19 October 1990,* 267–73. Los Angeles: Getty Conservation Institute. Reviews ethyl silicate reactions with earthen materials. Highlights the use of an ethyl silicate, Silester ZNS, in an ethanol and hydrochloric acid solution. Treatment did not change the appearance but imparted water resistance. Treated areas showed good erosion resistance.

Coffman, R., N. Agnew, G. Austin, and E. Doehne. Adobe mineralogy: Characterization of adobes from around the world. In *Proceedings from the Sixth International Conference on the Conservation of Earthen Architecture: Adobe 90 Preprints, Las Cruces, New Mexico, U.S.A., 14–19 October 1990,* 424–29. Los Angeles: Getty Conservation Institute. Compared *Conservare* OH and isocyanate resin for the consolidation of adobe. Major differences in consolidation are reported for adobe samples made in the laboratory versus original samples. Original pieces have more pore space; therefore, the consolidant does not act as effectively.

Domaslowski, W., D. Sobkowiak, and J. Wiklendt. The use of silicic acid hydrosols in stone consolidation. *Zabytkoznawstwo i Konservatorstwo* 14:23–62. [Polish with English summary] Applied silicic acid sols to sandstone and limestone.

Esbert, R. M., C. M. Grossi, L. Valdeón, J. Ordaz, and F. J. Alonso. Studies for stone conservation at the Cathedral of Murcia (Spain). In *Conservation of Monuments in the Mediterranean Basin,* ed. F. Zezza, 437–41. Brescia: Grafo Edizioni, 1990. Lab tests on limestones treated with ethyl silicate, polysiloxanes, and polyurethane.

Esbert, R. M., C. M. Grossi, L. Valdeon, J. Ordaz, and F. J. Alonso. Estudios de laboratorio sobre la conservación de la piedra de la Catedral Murcia. *Materiales de Construcción* 40(217):5–15. Spanish-language version of previous abstract.

Fassina, V. Considerazioni sui criteri di scelta del consolidante e la relative metodologia di applicazione. In *Il Prato della Valle e le opere in pietra calcarea collocate all'aperto: Esperienze e metodologie di conservazione in area Veneta. Atti della giornata di studio: Padova, 6 aprile 1990*, ed. S. Borsella, V. Fassina, and A. M. Spiazzi, 131–45. Padova: Libreria Progetto Editore, 1990. [Italian] Use of "silicone resin" to consolidate decayed limestone statues.

Helmi, F. Deterioration and conservation of some mud-brick in Egypt. In *Proceedings from the Sixth International Conference on the Conservation of Earthen Architecture: Adobe 90 Preprints, Las Cruces, New Mexico, U.S.A., 14–19 October 1990*, 277–82. Los Angeles: Getty Conservation Institute. Tested ethyl silicate and MTMOS on mud-brick.

Huang Kezhog, J. Huaiying, C. Run, and F. Lijuan. The weathering characteristics of the rocks of the Kezier Grottoes and research into their conservation. In *Proceedings from the Sixth International Conference on the Conservation of Earthen Architecture: Adobe 90 Preprints, Las Cruces, New Mexico, U.S.A., 14–19 October 1990*, 283–88. Los Angeles: Getty Conservation Institute. Used methyltriethoxysilane in conjunction with potassium silicate and magnesium fluorosilicate in an attempt to formulate a product that combined the best properties of inorganic and organic constituents.

Laurenzi Tabasso, M. Riflessioni sui problemi di conservazione dei materiali porosi da costruzione. In *Conservation of Monuments in the Mediterranean Basin*, ed. F. Zezza, 427–35. Brescia: Grafo Edizioni, 1990. General report on stone conservation including consolidation.

Marchesini, L. Tecnologie e modalita di intervento nel restauro della pietra tenera. In *Il Prato della Valle e le opere in pietra calcarea collocate all'aperto: Esperienze e metodologie di conservazione in area Veneta. Atti della giornata di studio: Padova, 6 aprile 1990*, ed. S. Borsella, V. Fassina, and A. M. Spiazzi, 65–71. Padova: Libreria Progetto Editore, 1990. [Italian] Report on the use of acrylic-silicone resin to treat soft limestone.

Mayer, H., and M. Roth. Silicon-Microemulsions-Konzentrate: Wassrige Bautenschutzmittel auf Basis silicium organischer Verbindungen. *Bautenschutz und Bausanierung* 13(1):1–4. [German] Describes new product, water-based microemulsions of alkoxysilane and siloxane, with acetic acid catalyst.

Muñoz, E. G., and M. P. Bahamóndez. Conservación de un sitio arqueológico construido en tierra. In *Proceedings from the Sixth International Conference on the Conservation of Earthen Architecture: Adobe 90 Preprints, Las Cruces, New Mexico, U.S.A., 14–19 October 1990*, 371–76. Los Angeles: Getty Conservation Institute. Use of ethyl silicate–based consolidants for the treatment of earthen walls.

Pietropoli, F. Il restauro del protiro de Duomo di Verona. In *Il Prato della Valle e le opere in pietra calcarea collocate all'aperto: Esperienze e metodologie di conservazione in area Veneta. Atti della giornata di studio: Padova, 6 aprile 1990*, ed. S. Borsella, V. Fassina, and A. M. Spiazzi, 169–73. Padova: Libreria Progetto Editore, 1990. [Italian] Decayed zones of limestone were consolidated with ethyl silicate.

Sattler, L., E. Wendler, R. Snethlage, and D. D. Klemm. Konservierung von carbonatisiertem Grunsandstein an der Alte Pinakothek in München. *Bautenschutz und Bausanierung* 13(6):93–97. Examines effects achieved by using different organosilicon compounds.

Selwitz, C., R. Coffman, and N. Agnew. The Getty Adobe Research Project at Fort Selden III: An evaluation of the application of chemical consolidants to test walls. In *Proceedings from the Sixth International Conference on the Conservation of Earthen Architecture: Adobe 90 Preprints, Las Cruces, New Mexico, U.S.A., 14–19 October 1990*, 255–60. Los Angeles: Getty Conservation Institute. Compared isocyanate resin with *Conservare* H and found that aged adobe structures took up both consolidants but were not mechanically strengthened.

Šrámek, J., and L. Losos. Outline of mud-brick structures conservation at Abusir, Egypt. In *Proceedings from the Sixth International Conference on the Conservation of Earthen Architecture: Adobe 90 Preprints, Las Cruces, New Mexico, U.S.A., 14–19 October 1990*, 449–54. Los Angeles: Getty Conservation Institute. Alkoxysilanes were used in conjunction with acrylics to give hydrophobation of the surface. ACRYLOID B72 used for deep impregnation. Alkoxysilanes (Wacker OH) were mixed with water prior to application, which was one week after B72 consolidation. Surfaces reported hydrophobic after one month.

Wheeler, G., and S. A. Fleming. Modulus of rupture for room temperature gels derived from methyltrimethoxysilane. In *The Engineering Geology of Ancient Works, Monuments and Historic Sites: Proceedings of an International Symposium Organized by the Greek National Group of IAEG, Athens, 19–23 September 1988*, vol. 4, ed. P. G. Marinos and G. C. Koukis, 2083–85. Rotterdam: A. A. Balkema, 1990. Determination of the strength of alkoxysilane gels from MTMOS, ETMOS, TEOS, and TMOS by three-point bend testing.

1991

Alessandrini, G., R. Peruzzi, S. Righini Ponticelli, T. De Dominicis, and L. Formica. La Basilica di S. Michele in Pavia: Problemi conservativi. In *Le Pietre nell'architettura: Struttura e superfici. Atti del convegno di studi Bressanone, 25–28 giugno 1991*, Scienza e beni culturali 7, ed. G. Biscontin and D. Mietto, 693–703. Padova: Libreria Progetto Editore, 1991. Test results suggest ethyl silicate and polysiloxane together are the best treatment for sandstone.

Bell, F. G. Preservation of stonework and a review of some methods used in the UK. *Architectural Science Review* 34(4):133–38. General review of preservation methods with mention of the use of alkoxysilanes on the west façade of Wells Cathedral.

Butlin, R. N., A. T. Coote, K. D. Ross, and T. J. S. Yates. Weathering and conservation studies at Wells Cathedral, England. In *Science, Technology and European Cultural Heritage: Proceedings of the European Symposium, Bologna, Italy, 13–16 June 1989*, ed. N. S. Baer, C. Sabbioni, and A. Sors, 306–9. Oxford: Butterworth-Heinemann, 1991. Reports on test of various consolidants (including alkoxysilanes) on limestone samples exposed at the Wells Cathedral site.

Butlin, R. N., T. J. S. Yates, J. R. Ridal, and D. Bigland. Studies of the use of preservative treatments on historic buildings. In *Science, Technology and European Cultural Heritage: Proceedings of the European Symposium, Bologna, Italy, 13–16 June 1989*, ed. N. S. Baer, C. Sabbioni, and A. Sors, 664–67. Oxford: Butterworth-Heinemann, 1991. Results of in situ weathering of 14 treatments, including alkoxysilanes.

Coffman, R. L., N. Agnew, and C. Selwitz. Modification of the physical properties of natural and artificial adobe by chemical consolidation. In *Materials Issues in Art and Archaeology II: Symposium Held 17–21 April 1990, San Francisco, California, U.S.A.*, ed. P. B. Vandiver, J. Druzik, and G. Wheeler, 201–7. Pittsburgh: Materials Research Society, 1991. Tested isocyanate resin *Conservare* H and OH on artificial adobe. Both the alkoxysilanes and the isocyanate were found to slow down deterioration due to exposure to water, but, overall, the isocyanate resin gave greater compressive strength increases.

Cosentino, M., F. Terranova, G. Margiotta, N. Doria, L. Pellegrino, and F. Mannuccia. Restauro conservativo del prospetto lapideo della chiesa del collegio dei gesuiti di Trapani. In *La Pietre nell'architettura: Struttura e superfici. Atti del convegno di studi Bressanone, 25–28 giugno 1991*, Scienza e beni culturali 7, ed. G. Biscontin and D. Mietto, 732–37. Padova: Libreria Progetto Editore, 1991. Reports on the use of ethyl silicates to consolidate limestone, marble, and travertine.

Esbert, R. M., C. Grossi, J. Ordaz, and F. J. Alonso. La conservación de la piedra de la Casa Milá ("La Pedrera" de Gaudí, Barcelona): Pruebas preliminarios. *Boletín Geológico y Minero* 102(3):446–54. Tested Dri Film 104, Wacker OH, and Tegovakon V on Vilafranca stone.

Esbert, R. M., M. Montoto, L.M. Suárez del Rio, V. G. Ruiz de Argandoña, and C. M. Grossi. Mechanical stresses generated by crystallization of salts inside treated and nontreated monumental stones: Monitoring and interpretation by acoustic emission/microseismic activity. In *Materials Issues in Art and Archaeology II: Symposium Held 17–21 April 1990, San Francisco, California, U.S.A.*, ed. P. B. Vandiver, J. Druzik, and G. Wheeler, 285–96. Pittsburgh: Materials Research Society, 1991. Limestone treated with Tegovakon V and water repellent. Preliminary results show that the treated samples give different results than do untreated samples.

Häberl, K., A. Rademacher, and G. Grassegger. Design of models and finite-element-calculation for load distribution in sedimentary rocks including the influence of silicic acid strengtheners. In *Durability of Building Materials and Components: Proceedings of the Fifth International Conference, Brighton, U.K., 7–9 November 1990*, ed. J. M. Baker, P. J. Nixon, A. J. Majumdar, and H. Davies. New York: E. & F. N. Spon, 1991. Proposes a model of different grain structures to determine the appropriate consolidant with emphasis on alkoxysilanes.

Honsinger, D., and H. R. Sasse. Alteration of microstructure and moisture characteristics of stone materials due to impregnation. In *Durability of Building Materials and Components: Proceedings of the Fifth International Conference, Brighton, U.K., 7–9 November 1990*, ed. J. M. Baker, P. J. Nixon, A. J. Majumdar, and H. Davies, 213–25. New York: E. & F. N. Spon, 1991. Examines alteration in pore structure of several stones with seven consolidants, including a modified ethyl silicate.

Leyden, D. E., and J. B. Atwater. Hydrolysis and condensation of alkoxysilanes investigated by internal reflection FTIR spectroscopy. *Journal of Adhesion Science Technology* 5(10):815–29. Studied hydrolysis and condensation of trimethyl-methoxysilane and ethyltrimethoxysilane in acidified water and acetone with FTIR. Found a sequential hydrolysis of alkoxy groups followed by the condensation of the resulting silanol groups. The acid catalysis was less effective as the alkyl groups became larger.

Leznicka, S., J. Kuroczkin, W. E. Krumbein, A. B. Strzelczyk, and K. Petersen. Studies on the growth of selected fungal strains on limestone impregnated with silicone resins (Steinfestiger H and Elastosil E-41). *International Biodeterioration* 28(1–4):91–111. Consolidation products were found not to inhibit fungal growth.

Pernice, F., and A. Rava. Restauro della facciata della chiesa della missione a Mondovi Cuneo. In *Le Pietre nell'architettura: Struttura e superfici. Atti del convegno di studi Bressanone, 25–28 giugno 1991*, Scienza e beni culturali 7, ed. G. Biscontin and D. Mietto, 455–62. Padova: Libreria Progetto Editore, 1991. Used Wacker OH to consolidate sandstone.

Plueddemann, E. *Silane Coupling Agents.* 2d ed. New York: Plenum, 1991. Fundamental text on the use of alkoxysilane coupling agents for composite materials.

Rava, A. Metodologia di intervento per il restauro delle arenarie di facciata della chiesa abbaziale di Vezzolano. In *Le Pietre nell'architettura: Struttura e superfici. Atti del convegno di studi Bressanone, 25–28 giugno 1991,* Scienza e beni culturali 7, ed. G. Biscontin and D. Mietto, 442–46. Padova: Libreria Progetto Editore, 1991. Used Wacker OH to consolidate sandstone.

Rocchi, P., and C. Piccirilli. *Manuale del consolidamento.* Rome: Tipografia del Genio Civile, DEI, 1991. Technical handbook on the consolidation of buildings.

Ruggieri, G., E. Cajano, G. Delfini, P. Mora, L. Mora, and G. Torraca. Il restauro conservativo della facciata di S. Andrea della valle in Roma. In *Le Pietre nell'architettura: Struttura e superfici. Atti del convegno di studi Bressanone, 25–28 giugno 1991,* Scienza e beni culturali 7, ed. G. Biscontin and D. Mietto, 535–44. Padova: Libreria Progetto Editore, 1991. Reports on the breakdown of ethyl silicate–derived gels on travertine.

Snethlage, R., and E. Wendler. Surfactants and adherent silicon resins: New protective agents for natural stone. In *Materials Issues in Art and Archaeology II: Symposium Held 17–21 April 1990, San Francisco, California, U.S.A.,* ed. P. B. Vandiver, J. Druzik, and G. Wheeler, 193–200. Pittsburgh: Materials Research Society, 1991. Describes modification of ethyl silicates by the introduction of elastic "bridges" to minimize shrinkage and cracking.

Vergès-Belmin, V., G. Orial, D. Garnier, A. Bouineau, and R. Coignard. Impregnation of badly decayed Carrara marble by consolidating agents: Comparison of seven treatments. In *Conservation of Monuments in the Mediterranean Basin: Proceedings of the Second International Symposium, Genève, 19–21 novembre 1991,* ed. D. Decrouez, J. Chamay, and F. Zezza, 421–37. Geneva: Ville de Genève, Muséum d'Histoire naturelle & Musée d'art et d'histoire, 1991. Badly decayed marble was consolidated with a number of products: ethyl silicate (Wacker OH), ethyl silicate followed by methylphenylpolysiloxane (MPPS–RC 11309), MPPS alone, a mixture of MPPS and ethyl silicate, ACRYLOID B72, and a fluorinated copolymer (Fomblin CO). The samples were consolidated, subjected to artificial weathering, and evaluated by taking sonic velocity measurements before and after weathering. Ethyl silicate (applied three times) and the MPPS gave the best results. The MPPS turned the marble yellow, which was attributed to the solubilization of organic matter in the black crust by toluene.

Wendler, E., D. Klemm, and R. Snethlage. Consolidation and hydrophic treatments of natural stone. In *Durability of Building Materials and Components: Proceedings of the Fifth International Conference, Brighton, U.K., 7–9 November 1990,* ed. J. M. Baker, P. J. Nixon, A. J. Majumdar, and H. Davies, 203–12. New York: E. & F. N. Spon, 1991. Describes modification of ethyl silicates by the introduction of elastic "bridges" to minimize shrinkage and cracking.

Wheeler, G., S. A. Fleming, and S. Ebersole. Evaluation of some current treatments for marble. In *Conservation of Monuments in the Mediterranean Basin: Proceedings of the Second International Symposium, Genève, 19–21 novembre 1991,* ed. D. Decrouez, J. Chamay, and F. Zezza, 439–43. Geneva: Ville de Genève, Muséum d'Histoire naturelle & Musée d'art et d'histoire, 1991. Assesses the performances of alkoxysilane (including coupling agents) and organic resin consolidants on marble by three-point bend testing.

Wheeler, G., G. Shearer, S. Fleming, L. W. Kelts, A. Vega, and R. J. Koestler. Toward a better understanding of B72 acrylic resin/methyltrimethoxysilane stone consolidants. In *Materials Issues in Art and Archaeology II: Symposium Held 17–21 April 1990, San Francisco, California, U.S.A.*, ed. P. B. Vandiver, J. Druzik, and G. Wheeler, 209–26. Pittsburgh: Materials Research Society, 1991. The addition of ACRYLOID B72 did not slow the evaporation of MTMOS but increased the mass retention of the alkoxysilane. Evaporation of MTMOS was greater on limestone versus sandstone samples with or without B72. In addition, B72 in toluene was found to consolidate limestone as well as or better than B72/MTMOS.

1992

F. G. Bell, The durability of some sandstones used in the United Kingdom as building stone, with a note on their preservation. In *Proceedings of the Seventh International Congress on Deterioration and Conservation of Stone, Held in Lisbon, Portugal, 15–18 June 1992*, ed. J. Delgado Rodrigues, F. Henriques, and F. Telmo Jeremias, 875–84. Lisbon: LNEC, 1992. Reviews sandstone weathering and treatments.

Danehey, C., G. Wheeler, and S. H. Su. The influence of quartz and calcite on the polymerization of methyltrimethoxysilane. In *Proceedings of the Seventh International Congress on Deterioration and Conservation of Stone, Held in Lisbon, Portugal, 15–18 June 1992*, ed. J. Delgado Rodrigues, F. Henriques, and F. Telmo Jeremias, 1043–52. Lisbon: LNEC, 1992. Solutions of water, MTMOS, and methanol in contact with quartz or calcite powders were studied by NMR. Calcite was shown to slow the condensation reaction.

De Witte, E., and K. Bos. Conservation of ferruginous sandstone used in northern Belgium. In *Proceedings of the Seventh International Congress on Deterioration and Conservation of Stone, Held in Lisbon, Portugal, 15–18 June 1992*, ed. J. Delgado Rodrigues, F. Henriques, and F. Telmo Jeremias, 1113–21. Lisbon: LNEC, 1992. Ferruginous sandstone was treated with Wacker OH and Tegovakon V. The treated samples were subjected to freeze-thaw and salt crystallization. Little damage was noted for SO_2, and consolidated samples resisted salt and freeze-thaw tests except when sulfates were present.

Dias, G. P. Stone conservation: Cleaning and consolidation. In *Proceedings of the Seventh International Congress on Deterioration and Conservation of Stone, Held in Lisbon, Portugal, 15–18 June 1992*, ed. J. Delgado Rodrigues, F. Henriques, and F. Telmo Jeremias, 1263–71. Lisbon: LNEC, 1992. General review of cleaning and consolidation methods and materials.

Elfving, P., and U. Jäglid. Silane bonding to various mineral surfaces. Report OOK 92:01, ISSN 0283-8575. Götenborg: Department of Inorganic Chemistry, Chalmers University of Technology, 1992. FTIR was used to evaluate the chemical bonding of trimethylmethoxysilane to various minerals. The spectra of treated silicate minerals indicated bonding that was absent for treated samples of calcite and gypsum.

Hammecker, C., R. M. E. Alemany, and D. Jeannette. Geometry modifications of porous networks in carbonate rocks by ethyl silicate treatment. In *Proceedings of the Seventh International Congress on Deterioration and Conservation of Stone, Held in Lisbon, Portugal, 15–18 June 1992*, ed. J. Delgado Rodrigues, F. Henriques, and F. Telmo Jeremias, 1053–62. Lisbon: LNEC, 1992. RC70 (ethyl silicate) and RC80 (ethyl silicate and dimethylsilicone resin) were tested on two limestones, Laspra and Hontoria. The Laspra stone (greater porosity, finer texture, larger specific surface area) had a more homogeneous distribution of the RC70 than did the Hontoria.

The RC80 had the reverse distribution: silicone resin was concentrated on the surface of the Laspra samples, and the Hontoria had a more homogeneous distribution in depth.

Hosek, J., and J. Šrámek. Arkoses and their highly hydrophobic treatments. In *Proceedings of the Seventh International Congress on Deterioration and Conservation of Stone, Held in Lisbon, Portugal, 15–18 June 1992*, ed. J. Delgado Rodrigues, F. Henriques, and F. Telmo Jeremias, 1197–1204. Lisbon: LNEC, 1992. Samples treated with alkyl-lalkoxysilanes and acrylic resins were found to be more sensitive to water adsorption.

Miller, E. Current practice at the British Museum for the consolidation of decayed porous stones. *The Conservator* 16:78–84. General review of alkoxysilane-based consolidants used at the British Museum, with practical descriptions of techniques.

Saleh, S. A., F. M. Helmi, M. M. Kamal, and A. E. El-Banna. Artificial weathering of treated limestone: Sphinx, Giza, Egypt. In *Proceedings of the Seventh International Congress on Deterioration and Conservation of Stone, Held in Lisbon, Portugal, 15–18 June 1992*, ed. J. Delgado Rodrigues, F. Henriques, and F. Telmo Jeremias, 781–89. Lisbon: LNEC, 1992. Applied several consolidants—5 % w/v ACRYLOID B72 in Hey'di M.S. Siloxane, MTMOS, 2.5% w/v ACRYLOID B72 in MTMOS, Wacker OH—to salt-contaminated limestone and exposed them to 20% solutions of HCl, UV radiation, relative humidity cycling, and bacterial growth. MTMOS solution was shown to give the best overall performance.

Saleh, S. A., F.M. Helmi, M. M. Kamal, and A. E. El-Banna. Study and consolidation of sandstone: Temple of Karnak, Luxor, Egypt. *Studies in Conservation* 37(2):93–104. Same consolidants and samples as previous abstract. ACRYLOID B72/MTMOS and Wacker OH gave the highest compressive strengths; ACRYLOID B72/Hey'di M.S. Siloxane and ACRYLOID B72/MTMOS gave the largest increases in tensile strengths, and Wacker OH the lowest.

Saleh, S. A., F. M. Helmi, M. M. Kamal, and A. E. El-Banna. Consolidation study of limestone: Sphinx, Giza, Egypt. In *Proceedings of the Seventh International Congress on Deterioration and Conservation of Stone, Held in Lisbon, Portugal, 15–18 June 1992*, ed. J. Delgado Rodrigues, F. Henriques, and F. Telmo Jeremias, 791–800. Lisbon: LNEC, 1992. Analysis of the limestone showed a mixture of calcite, quartz, gypsum, anhydrite, halite, goethite, and clay minerals. Using the same consolidants listed in the previous two abstracts, the authors concluded that MTMOS gave the best results: decreased water absorption by 82%, increased compressive strength by 124%, increased tensile strength by 16.8%.

Sattler, L. Untersuchungen zu Wirkung und Dauerhaftigkeit von Sandsteinfestigungen mit Kieselsäureester. Doctoral thesis, Ludwig-Maximilian Univeritat (Munich), 1992. Fundamental work on the use of ethyl silicate for the consolidation of sandstone.

Selwitz, C. *Epoxy Resins in Stone Conservation*. Research in Conservation Series, vol. 7. Marina del Rey, Calif.: Getty Conservation Institute, 1992. Emphasizes the large improvements in mechanical strengths from epoxy resins as compared to silicon derivatives.

Simon, S., H. P. Boehm, and R. Snethlage. A surface-chemical approach to marble conservation. In *Proceedings of the Seventh International Congress on Deterioration and Conservation of Stone, Held in Lisbon, Portugal, 15–18 June 1992*, ed. J. Delgado Rodrigues, F. Henriques, and F. Telmo Jeremias, 851–59. Lisbon: LNEC, 1992. Used quaternary ammonium polydimethylsiloxane on calcite to reduce dissolution in water.

Wendler, E., L. Sattler, P. Zimmermann, D. D. Klemm, and R. Snethlage. Protective treatment of natural stone: Requirements and limitations with respect to the state of damage. In *Proceedings of the Seventh International Congress on Deterioration and Conservation of Stone, Held in Lisbon, Portugal, 15–18 June 1992*, ed. J. Delgado Rodrigues, F. Henriques, and F. Telmo Jeremias, 1103–12. Lisbon: LNEC, 1992. Ethyl silicates increased the hygric dilation of a clay-rich sandstone.

Wheeler, G., S. A. Fleming, and S. Ebersole, Comparative strengthening effect of several consolidants on Wallace sandstone and Indiana limestone. In *Proceedings of the Seventh International Congress on Deterioration and Conservation of Stone, Held in Lisbon, Portugal, 15–18 June 1992*, ed. J. Delgado Rodrigues, F. Henriques, and F. Telmo Jeremias, 1033–41. Lisbon: LNEC, 1992. Several treatments (including Wacker H and OH, MTMOS, and alkoxysilane coupling agents) on sandstone and limestone were evaluated by three-point bend testing. Coupling agents improved strength increases.

Wheeler, G., A. Schein, G. Shearer, S. H. Su, and C. S. Blackwell. Preserving our heritage in stone. *Analytical Chemistry* 64(5):347–56. Review of stone consolidation focusing on alkoxysilanes.

Wheeler, G., E. Wolkow, and H. Gafney. Microstructures of B72 acrylic resin/MTMOS composites. In *Materials Issues in Art and Archaeology III: Symposium Held April 27–May 1, 1992, San Francisco, California, U.S.A.*, ed. P. B. Vandiver, J. R. Druzik, G. Wheeler, and I. C. Freestone, 963–67. Pittsburgh: Materials Research Society, 1992. Examined fracture surfaces of monoliths prepared from neat MTMOS and MTMOS/ACRYLOID B72 by SEM. Neat MTMOS shows a glasslike fracture, whereas ACRYLOID B72/MTMOS shows spheres of MTMOS-derived gel in an acrylic matrix.

Zanardi, B., L. Calzetti, A. Casoli, A. Mangia, G. Rizzi, and S. Volta. Observations on a physical treatment of stone surfaces. In *Proceedings of the Seventh International Congress on Deterioration and Conservation of Stone, Held in Lisbon, Portugal, 15–18 June 1992*, ed. J. Delgado Rodrigues, F. Henriques, and F. Telmo Jeremias, 1243–51. Lisbon: LNEC, 1992. The consolidation treatment consisted of three steps: preconsolidation with Wacker OH, water washing, application of Wacker OH.

1993

Biscontin, G., P. Maravelaki, E. Zendri, A. Glisenti, and E. Tondello. Investigation into the interaction between aqueous and organic solvent protection and building materials. In *Conservation of Stone and Other Materials: Proceedings of the International RILEM/UNESCO Congress "Conservation of Stone and Other Materials—Research-Industry-Media," Held at UNESCO Headquarters, Paris, with the Cooperation of ICCROM, Paris, June 19–July 1, 1993, II*, ed. M.-J. Thiel, 689–96. New York: E. & F. N. Spon, 1993. Applied 10% solutions of octyltriethoxysilane in white spirits and propyltrimethoxysilane in water to artificially weathered Vicenza stone and brick. Both the stone and the brick absorbed more of the propyltrimethoxysilane. The presence of NaCl increased water absorption by 200–300% for treated samples.

Capponi, G., and M. Laurenzi Tabasso. Esperienze italiane per la conservazione delle superfici lapidee dell'architettura: L'essempio del barocco leccese. In *Alteración de granitos y rocas afines, empleados como materiales de construcción: Deterioro de monumentos históricos. Actas del workshop, Consejo Superior de Investigaciones Científicas, Ávila, Spain, 1993*, ed. M. A. Vicente Hernandez, E. Molina Ballesteros, and V. Rives Arnau, 51–58. Rome: Istituto Centrale del Restauro, 1993. Outlines developments in the conservation of architectural stone surfaces in Italy including the selection of consolidants.

De Casa, G., M. Laurenzi Tabasso, and U. Santamaria. Study for consolidation and protection of Proconnesian marble. In *Conservation of Stone and Other Materials: Proceedings of the International RILEM/UNESCO Congress "Conservation of Stone and Other Materials—Research-Industry-Media," Held at UNESCO Headquarters, Paris, with the Cooperation of ICCROM, Paris, June 19–July 1, 1993, II,* ed. M.-J. Thiel, 768–74. New York: E. & F. N. Spon, 1993. Subjected samples of Proconnesian marble treated with RC80, RC90, Rhodorsil 10336 to artificial aging. RC80 and RC90 had greater depth of penetration. Rhodorsil 10336 samples had less weight loss. Water repellency for RC90 decreased with weathering and increased for RC80. RC90 samples were found to yellow strongly.

Esbert, R. M., and F. Diaz-Pache. Influencia de las caracteristicas petrofisicas en rocas monumentales porosas. *Materiales de Construcción* 43(230):25–36. Penetration of ethyl silicate consolidants is influenced by petrophysical properties of the rock.

Leznicka, S., and J. Lukaszewicz. The growth of fungi and algae on limestone and sandstone impregnated with organosilicates. In *Naukowe podstawy ochrony i konserwacji dzie sztuki oraz zabytków kultury materialnej,* ed. A. Strzelczyk and S. Skibinski, 211–21. Torun: Nicholas Copernicus University, 1993. Silicone resins generally were found to increase biological growth.

Sasse, H. R., D. Honsinger, and B. Schwamborn. PINS: A new technology in porous stone conservation. In *Conservation of Stone and Other Materials: Proceedings of the International RILEM/UNESCO Congress "Conservation of Stone and Other Materials—Research-Industry-Media," Held at UNESCO Headquarters, Paris, with the Cooperation of ICCROM, Paris, June 19–July 1, 1993, II,* ed. M.-J. Thiel, 705–16. New York: E. & F. N. Spon, 1993. Polymer Impregnated Natural Stone (PINS) refers to techniques used for checking the effectiveness of consolidating treatments. Two hundred polymeric products were tested, including organosilicone-modified polyurethanes, epoxy resins, acrylic resins, organosilicone compounds (alkoxysilanes and polysiloxanes), fluorocarbons, and unsaturated polyesters.

1994

Alonso, F. J., R. M. Esbert, J. Alonso, and J. Ordaz. Saline spray action on a treated dolomitic stone. In *The Conservation of Monuments in the Mediterranean Basin: Proceedings of the Third International Symposium, Venice, 22–25 June 1994,* ed. V. Fassina, H. Ott, and F. Zezza, 867–70. Venice: Soprintendenza ai Beni Artistici e Storici di Venezia, 1994. Studies the combination of a Tegovakon product and a silicone water repellent on a dolomitic limestone.

Bahamondez Prieto, M. Acciones de conservación sobre los moai de Isla de Pascua: Su evaluacion en laboratorio. In *Lavas and Volcanic Tuffs: Proceedings of the International Meeting, Easter Island, Chile, 25–31 October 1990,* ed. A. E. Charola, R. J. Koestler, and G. Lombardi, 89–99. Rome: ICCROM, 1994. Ethyl silicates shown to be an effective consolidant for volcanic tuff.

Berry, J. The encapsulation of salts by consolidants used in stone conservation. *Institute of Archaeology Papers,* no. 5, 1994, 29–37. Uses SEM to examine MTMOS, Wacker H and OH gels on salt crystals.

Berry, J., and C. A. Price. The movement of salts in consolidated stone. In *The Conservation of Monuments in the Mediterranean Basin: Proceedings of the Third International Symposium, Venice, 22–25 June 1994,* ed. V. Fassina, H. Ott, and F. Zezza, 845–48. Venice: Soprintendenza ai Beni Artistici e Storici di Venezia, 1994. SEM demonstrated that Wacker H and OH gels did not form impenetrable barriers around salt crystals.

Elfving, P., L.-G. Johansson, and O. Lindqvist. A study of the sulphation of silane-treated sandstone and limestone in a sulphur dioxide atmosphere. *Studies in Conservation* 39(3):199–209. SO$_2$ deposition is reduced by the ethyl silicate/alkyltriethoxysilane treatments of sandstone and limestone.

Félix, C. Déformation de grès consecutives à leur consolidation avec un silicate d'éthyle. In *7th International Congress of the Association of Engineering Geology, Lisboa,* ed. R. Oliveria, L. F. Rodrigues, A. G. Coehlo, and A. P. Cunha, 3543–50. Rotterdam: A. A. Balkema, 1994. Examined the shrinkage of sandstone due to consolidation with ethyl silicate.

Félix, C. and V. Furlan. Variations dimensionnelles de grès et calcaires, liées à leur consolidation avec un silicate d'éthyle. In *The Conservation of Monuments in the Mediterranean Basin: Proceedings of the Third International Symposium, Venice, 22–25 June 1994,* ed. V. Fassina, H. Ott, and F. Zezza, 855–59. Venice: Soprintendenza ai Beni Artistici e Storici di Venezia, 1994. Study of shrinkage and swelling sandstone treated with ethyl silicates.

Galan, E., and M. I. Carretero. Estimation of the efficacy of conservation treatments applied to a permotriassic sandstone In *The Conservation of Monuments in the Mediterranean Basin: Proceedings of the Third International Symposium, Venice, 22–25 June 1994,* ed. V. Fassina, H. Ott, and F. Zezza, 947–54. Venice: Soprintendenza ai Beni Artistici e Storici di Venezia, 1994. Dolomitic sandstone treated with Tegovakon V was subjected to accelerated/artificial weathering. The consolidant initially slowed deterioration (versus treated samples), but damage was the same for both by the end of testing.

Glossner, S. Untersuchungen zur Verklebung von Schuppen und Schalen an Naturstein mit schnellhydrolysiertem Kieselsäureester. *Arbeitsblätter für Restauratoren: Gruppe 6 – Stein* 27(1):289–94. [German with English summary] Different solvents were tested with TEOS. Samples were preconsolidated with ethanol, TEOS, and Plastorit. This treatment was followed by limewash and ethanol.

Goins, E. S. The acid–base surface characterization of sandstone, limestone, and marble and its effect upon the polymerisation of tetraethoxysilane. *Institute of Archaeology Papers,* no. 5, 1994, 19–28. Studies the acid and base characteristics of stone powders and their effect on the reactions of tetraethoxysilane solutions.

Hilbert, G., and F. Janning. Voruntersuchungen zur Steinfestigung am Stephansdom zu Wien. *Bautenschutz und Bausanierung* 17(1):42–44. [German with English summary] Ethyl silicate was used to consolidate German monuments. Treatments evaluated by studying Young's modulus and profiles of resistance to drilling.

Koblischek, P. J. Die Konsolidierung von Naturstein. *Arbeitsblätter für Restauratoren: Gruppe 6 – Stein* 27(1):295–301. [German with English summary] Tested precondensed, uncatalyzed ethyl silicate consolidants.

Koblischek, P. J. Polymers in the renovation of buildings constructed of natural stone in the Mediterranean basin. In *The Conservation of Monuments in the Mediterranean Basin: Proceedings of the Third International Symposium, Venice, 22–25 June 1994,* ed. V. Fassina, H. Ott, and F. Zezza, 849–54. Venice: Soprintendenza ai Beni Artistici e Storici di Venezia, 1994. Review of a number of consolidants and their application, focusing on Motema products.

Kumar, R., and C. A. Price. The influence of salts on the hydrolysis and condensation of methyltrimethoxysilane. In *The Conservation of Monuments in the Mediterranean Basin: Proceedings of the Third International Symposium, Venice, 22–25 June 1994,* ed. V. Fassina, H. Ott, and F. Zezza, 861–65. Venice: Soprintendenza ai Beni Artistici e Storici di Venezia, 1994. Studies the rates of hydrolysis and condensation for solutions of MTMOS in contact with various salts: sodium chloride, sodium sulfate, sodium nitrate, magnesium sulfate, and magnesium chloride.

Lukaszewicz, J. W. The application of silicone products in the conservation of volcanic tuffs. In *Lavas and Volcanic Tuffs: Proceedings of the International Meeting, Easter Island, Chile, 25–31 October 1990*, ed. A. E. Charola, R. J. Koestler, and G. Lombardi, 191–202. Rome: ICCROM, 1994. Neat Wacker OH and diluted with ethanol (1:1) were applied to a volcanic tuff. Samples treated with neat Wacker OH had hygroscopicity higher than that of the untreated samples. Application of Wacker OH followed by water repellent provided less protection against water and water vapor than the water repellents applied alone.

Nishiura, T., M. Okabe, and N. Kuchitsu. Study on the conservation treatment of Irimizu Sanjusan Kannon: Cleaning and protective treatment of a marble Buddha image. *Hozon kagaku* 33: 67–72. [Japanese with English summary] Describes the treatment of marble statues in poor condition with a solution containing methyltriethoxysilane (Colcoat SS-101).

Rubio, L., E. Andrés, M.A. Bello, J. F. Vale, and M. Alcalde. Evaluation of the hygric characteristics of diverse Spanish stones after the application of different protective treatments. In *The Conservation of Monuments in the Mediterranean Basin: Proceedings of the Third International Symposium, Venice, 22–25 June 1994*, ed. V. Fassina, H. Ott, and F. Zezza, 871–75. Venice: Soprintendenza ai Beni Artistici e Storici di Venezia, 1994. Evaluated the hygric characteristics of dolomitic limestones treated with several alkoxysilane-based systems.

Šrámek, J., and T. Nishiura. Assessment, by radioactive labelling, of the efficiency of conservation treatments applied to volcanic tuffs. In *Lavas and Volcanic Tuffs: Proceedings of the International Meeting, Easter Island, Chile, 25–31 October 1990*, ed. A. E. Charola, R. J. Koestler, and G. Lombardi, 205–16. Rome: ICCROM, 1994. Volcanic tuffs were treated with Wacker H and Wacker OH and labelled with a radioactive tracer. After accelerated weathering the amount of tracer lost was determined.

Laurenzi Tabasso, M., A. M. Mecchi, and U. Santamaria. Interaction between volcanic tuff and products used for consolidation and waterproofing treatments. In *Lavas and Volcanic Tuffs: Proceedings of the International Meeting, Easter Island, Chile, 25–31 October 1990*, ed. A. E. Charola, R. J. Koestler, and G. Lombardi, 173–90. Rome: ICCROM, 1994. Tegovakon V was applied to volcanic tuffs and found to improve compressive strength and "durability."

Useche, L. A. Studies for the consolidation of the façade of the Church of Santo Domingo, Popayan, Colombia. In *Lavas and Volcanic Tuffs: Proceedings of the International Meeting, Easter Island, Chile, 25–31 October 1990*, ed. A. E. Charola, R. J. Koestler, and G. Lombardi, 165–72. Rome: ICCROM, 1994. Weathered volcanic tuff was treated with Wacker H and OH, ACRYLOID B72, and an epoxy resin. The Wacker products gave the best results: large weight increase, reduced porosity, good increases in compressive strength, reduced water absorption. A 2-step process consisting in preconsolidation with Wacker OH followed by treatment with Wacker H gave the best results.

Wheeler, G., and R. Newman. Analysis and treatment of a stone urn from the Imperial Hotel, Tokyo. In *Lavas and Volcanic Tuffs: Proceedings of the International Meeting, Easter Island, Chile, 25–31 October 1990*, ed. A. E. Charola, R. J.

Koestler, and G. Lombardi, 157–63. Rome: ICCROM, 1994. *Conservare* OH treatment darkened the stone for several weeks but then returned to its original color, and the surface was significantly less friable.

1995

Butlin, R. N., T. J. S. Yates, and W. Martin. Comparison of traditional and modern treatments for conserving stone. In *Methods of Evaluating Products for the Conservation of Porous Building Materials in Monuments: International Colloquium, Rome, 19–21 June 1995: Preprints*, ed. M. Laurenzi Tabasso, 111–19. Rome: ICCROM, 1995. Reviews field evaluations over a twenty-year period for monuments treated with Brethane.

Caselli, A., and D. Kagi. Methods used to evaluate the efficacy of consolidants on an Australian sandstone. In *Methods of Evaluating Products for the Conservation of Porous Building Materials in Monuments: International Colloquium, Rome, 19–21 June 1995: Preprints*, ed. M. Laurenzi Tabasso, 121–30. Rome: ICCROM, 1995. Evaluated alkoxysilane stone consolidants (Wacker OH, Wacker H, and Brethane) on sandstone. All products increased compressive strengths by about 100%. Wacker OH was the only product to penetrate through the weathered zone.

Durán-Suárez, A., J. Garcia-Beltrán, and J. Rodríguez-Gordillo. Colorimetric cataloguing of stone materials (biocalcarenite) and evaluation of the chromatic effects of different restoring materials. *Science of the Total Environment* 167:171–80. Reports color changes to a biocalcarenite with consolidation. Ethyl silicate and alkoxysilane monomers yield little change in color.

Félix, C. Peut-on consolider les grès tendres du Plateau Suisse avec le silicate d'éthyle? In *Preservation and Restoration of Cultural Heritage: Stone Materials, Air Pollution, Murals, Scientific Research Work, and Case Studies: Proceedings of the 1995 LCCP Congress, Montreux 24–29 September 1995*, ed. R. Pancella, 267–74. Lausanne: Ecole Polytechnique Federale de Lausanne, 1995. Ethyl silicate consolidants are not recommended for the treatment of "les grès tendres du Plateau Suisse" as they are problematic and irreversible.

Garcia Pascua, N., M. I. Sanchez de Rojas, and M. Frias. Study of porosity and physical properties as methods to establish the effectiveness of treatments used in two different Spanish stones: Limestone and sandstone. In *Methods of Evaluating Products for the Conservation of Porous Building Materials in Monuments: International Colloquium, Rome, 19–21 June 1995: Preprints*, ed. M. Laurenzi Tabasso, 147–61. Rome: ICCROM, 1995. Stone samples treated with an ethyl silicate (Minersil SH) showed decreases in median pore diameter and large decreases in overall porosity.

Ginell, W. S., R. Kumar, and E. Doehne. Conservation studies on limestone from the Maya site at Xunantunich, Belize. In *Materials Issues in Art and Archaeology IV: Symposium Held 16–21 May 1994, Cancun, Mexico*, ed. P. B. Vandiver, J. Druzik, J. L. G. Madrid, I. C. Freestone, and G. Wheeler, 813–21. Pittsburgh: Materials Research Society, 1995. Tested a number of consolidants and environments both in situ and in the laboratory. All samples showed fungal growth in outdoor shady areas. *Conservare* H increased erosion resistance.

Goins, E. S. Alkoxysilane stone consolidants: The effect of the stone substrate on the polymerization process. Doctoral thesis, University College London, University of London, 1995. Studies MTMOS, water, and ethanol solutions in contact with limestone, sandstone, marble, and sodium chloride. Found that the limestone slowed the hydrolysis reaction of the alkoxysilane sols and produced a different type of gel.

Goins, E. S., G. Wheeler, and S. A. Fleming. The influence of reaction parameters on the effectiveness of tetraethoxysilane-based stone consolidants. In *Methods of Evaluating Products for the Conservation of Porous Building Materials in Monuments: International Colloquium, Rome, 19–21 June 1995: Preprints,* ed. M. Laurenzi Tabasso, 259–74. Rome: ICCROM, 1995. Sandstone and limestone are consolidated with a tetraethoxysilane-derived sol diluted in a number of organic solvents and evaluated by three-point bend testing. Sandstone cores were affected by the acid-base properties of the solvent, and these effects were absent for limestone cores.

Guidetti, V., M. Matullo, and G. Pizzigoni. Methodologies for the study of the efficiency of stone reaggregant products on artificial samples. In *Preservation and Restoration of Cultural Heritage: Stone Materials, Air Pollution, Murals, Scientific Research Work, and Case Studies: Proceedings of the 1995 LCCP Congress, Montreux 24–29 September 1995,* ed. R. Pancella, 237–46. Lausanne: Ecole Polytechnique Federale de Lausanne, 1995. Prepared gypsum-sand monoliths and treated them with an ethyl silicate consolidant, a fluorinated elastomer, and a polyfluorourethane resin.

Hellbrügge, C. Praktische Erfahrungen mit Anböschmaterialien auf Polyurethan-Basis beim Baumberger Kalksandstein. *Arbeitsblätter für Restauratoren: Gruppe 6 – Stein* 28(2):333–39. Reports on testing Funcosil products on calcareous sandstone.

Koblischek, P. The consolidation of natural stone with a stone strengthener on the basis of poly-silicic-acid-ethyl-ester. In *Preservation and Restoration of Cultural Heritage: Stone Materials, Air Pollution, Murals, Scientific Research Work, and Case Studies: Proceedings of the 1995 LCCP Congress, Montreux 24–29 September 1995,* ed. R. Pancella, 261–65. Lausanne: Ecole Polytechnique Federale de Lausanne, 1995. Describes the properties of the ethyl silicate–based Motema products.

Kumar, R. Fourier transform infrared spectroscopic study of silane/stone interface. In *Materials Issues in Art and Archaeology IV: Symposium Held 16–21 May 1994, Cancun, Mexico,* ed. P. B. Vandiver, J. Druzik, J. L. G. Madrid, I. C. Freestone, and G. Wheeler, 341–47. Pittsburgh: Materials Research Society, 1995. Studies bond formation between mineral powders and alkoxysilanes.

Kumar, R., and W. S. Ginell. Evaluation of consolidants for stabilization of weak Maya limestone. In *Methods of Evaluating Products for the Conservation of Porous Building Materials in Monuments: International Colloquium, Rome, 19–21 June 1995: Preprints,* ed. M. Laurenzi Tabasso, 163–78. Rome: ICCROM, 1995. Similar to Ginell, Kumar, and Doehne 1991.

Littmann, K., and B. Riecken. Stone-protecting agents and their chemical behaviour under the influence of weathering. In *Methods of Evaluating Products for the Conservation of Porous Building Materials in Monuments: International Colloquium, Rome, 19–21 June 1995: Preprints,* ed. M. Laurenzi Tabasso, 349–57. Rome: ICCROM, 1995. Compared polyurethane, aliphatic diglycidalether + hexamethylenediamine, and ethyl silicate + oligosiloxanes under both artificial and natural weathering conditions on three sandstones and one limestone. Found that after one year of artificial weathering, the ethyl silicate solution had lost much of its hydrophobicity, particularly on the limestone.

Lukaszewicz, J. W., D. Kwiatkowski, and M. Klingspor. Consolidation of Gotland stone in monuments. In *Methods of Evaluating Products for the Conservation of Porous Building Materials in Monuments: International Colloquium, Rome, 19–21 June 1995: Preprints,* ed. M. Laurenzi Tabasso, 179–87. Rome: ICCROM, 1995. Comments on in situ aging of Wacker OH and Funcosil OH on Gotland sandstone. Slight darkening of the stone and hydrophobic effect on the surface are long-lasting. Freshly quarried and treated samples cured at 75% RH had no water repellency, while those cured at 50% did.

Nandiwada, A., and C. A. Price. Retreatment of consolidated stone. In *Processes of Urban Stone Decay: Proceedings of SWAPNET '95 Stone Weathering and Atmospheric Pollution, Belfast, 19–20 May 1995*, ed. B. J. Smith and P. A. Warke, 261–65. New York: Donhead, 1995. Describes alkoxysilanes applied in multiple treatments.

Nishiura, T. Experimental evaluation of stone consolidants used in Japan. In *Methods of Evaluating Products for the Conservation of Porous Building Materials in Monuments: International Colloquium, Rome, 19–21 June 1995: Preprints*, ed. M. Laurenzi Tabasso, 189–202. Rome: ICCROM, 1995. Tested several alkoxysilane consolidants. A methyltriethoxysilane solution (Colcoat SS-101) performed best in a "split" test and abrasion resistance. Wacker OH was found to have superior penetrating ability but did not have good cohesive strength.

Perez, J. L., R. Villegas, J. F. Vale, M. A. Bello, and M. Alcalde. Effects of consolidant and water-repellent treatments on the porosity and pore-size distribution of limestones. In *Methods of Evaluating Products for the Conservation of Porous Building Materials in Monuments: International Colloquium, Rome, 19–21 June 1995: Preprints*, ed. M. Laurenzi Tabasso, 203–11. Rome: ICCROM, 1995. Measures the pore-size distribution before and after consolidation for a calcitic sandstone, a limestone, and a dolomite treated with Wacker OH, Tegovakon V, Wacker BS 28 and 290 L, Tegosivin HL100, ACRYLOID B72, and Raccanello 55050. Calcitic sandstone experienced considerable decreases in all pore sizes with Wacker OH. Tegovakon V had no appreciable effect on the pore size distribution.

Rossi-Manaresi, R., A. Rattazzi, and L. Toniolo. Long-term effectiveness of treatments of sandstone. In *Methods of Evaluating Products for the Conservation of Porous Building Materials in Monuments: International Colloquium, Rome, 19–21 June 1995: Preprints*, ed. M. Laurenzi Tabasso, 225–44. Rome: ICCROM, 1995. Twenty years of in situ weathering were reported on a calcareous sandstone. Areas treated with ACRYLOID B72/Dri Film 104 looked much the same as it had immediately after treatment; with Rhodorsil XR-893 were uneven; and with Wacker H had white discolorations, and consolidated crusts were hard, but not attached to the surface. Water absorption tests showed that the B72/Dri Film areas had not decreased in hydrophobicity, and the Rhodorsil still had good hydrophobicity. The Wacker H areas had significantly decreased in hydrophobicity with age.

Schwamborn, B., and B. Riecken. Behaviour of impregnated natural stones after different weathering procedures. In *Proceedings of the First International Symposium "Surface Treatment of Building Materials with Water Repellent Agents," Delft, 9–10 November 1995*, vol. 26, 1–14. Delft: Delft University of Technology, 1995. Tested water repellents and consolidants based on siloxanes, polyurethanes, and ethyl silicate for water uptake and strength after weathering. Also published in *International Zeitschrift für Bauinstandsetzen* 2, no. 2(1996):101–16.

Wendler, E. Gesteinfestigung mit marktublichen und modifizierten Kieselsaureester produkten. In *Natursteinsanierung Bern*, Interacryl AG, 1995, 1–7. Discusses elastomer-modified ethyl silicate.

1996

Alaimo, R., R. Giarrusso, L. Lazzarini, F. Mannuccia, and P. Meli. The conservation problems of the theatre of Eraclea Minoa (Sicily). In *Proceedings of the Eighth International Congress on Deterioration and Conservation of Stone: Berlin, 30 September–4 October 1996*, ed. Joseph Riederer, 1085–95. Berlin: Möller Druck und Verlag, 1996. Ethyl silicate (Rankover) consolidation followed with silicone resin water repellent gave the best consolidation of a biocalcarenite. Tegovakon V and Wacker OH did not perform as well.

Antonova, H. Evaluation of protective treatments in laboratory and in situ. *Proceedings of the Eighth International Congress on Deterioration and Conservation of Stone: Berlin, 30 September–4 October 1996*, ed. Joseph Riederer, 1277–83. Berlin: Möller Druck und Verlag, 1996. Tested polysiloxanes on limestone at 70% and 35% RH. Changes in RH during deposition influence the final properties of the coating.

Boos, M., J. Grobe, K. Meise-Gresch, S. Tarlach, and H. Eckert. Alterungsmerkmale unterschiedlicher Steinfestiger auf Basis von Kieselesäureestern (KSE). In *Werkstoffwissenschaften und Bauinstandsetzen. Berichtsband zum Vierten Internationalen Kolloquium, Vol. 1*, ed. F. H. Wittmann and A. Gerdes, 551–62. Berlin: AEDIFICATIO, 1996. Monomeric and polymeric ethyl silicate consolidants have similar active substances that contain few silanols. Catalyzed formulations lead to substances with different hydrophobic properties.

Boos, M., J. Grobe, G. Hilbert, and J. Müller-Rochholz. Modified elastic silicic-acid ester applied on natural stone and tests of their efficiency. In *Proceedings of the Eighth International Congress on Deterioration and Conservation of Stone: Berlin, 30 September–4 October 1996*, ed. Joseph Riederer, 1179–85. Berlin: Möller Druck und Verlag, 1996. Defines favorable consolidation when the strength versus depth profile approaches that of unweathered stone. Consolidated a number of sandstones with Funcosil Stone Strengtheners 100, 300, and 510. Compared destructive and nondestructive (ultrasonic) methods for evaluating the quality of consolidation treatments.

Bruchertseifer, Chr., S. Brüggerhoff, J. Grobe, and H. J. Götze. DRIFT investigation of silylated natural stone—molecular surface information and macroscopic features. In *Proceedings of the Eighth International Congress on Deterioration and Conservation of Stone: Berlin, 30 September–4 October 1996*, ed. Joseph Riederer, 1223–31. Berlin: Möller Druck und Verlag, 1996. Uses diffuse reflectance FTIR (DRIFTS), in conjunction with hygric behavior, to study surfaces of two sandstones treated with several alkoxysilanes and silicone resins.

Brus, J., and P. Kotlík. Consolidation of stone by mixtures of alkoxysilane and acrylic polymer. *Studies in Conservation* 41(2):109–19. Uses a mixture of ethyl silicate 40 and ACRYLOID B72 to consolidate sandstone.

Brus, J., and P. Kotlík. Cracking of organosilicone stone consolidants in gel form. *Studies in Conservation* 41(1):55–59. Prepared monoliths from Wacker H and OH, Tegovakon V and T(?), and ethyl silicate 29 and then immersed them in a solvent (water, acetone, ethanol) to study their behavior. The stability of the gel was related to catalyst type. Gels prepared from acid catalyzed sols were much more stable than those prepared from dibutyltindilaurate catalysts.

Ciabach, J. The effect of water-soluble salts on the impregnation of sandstone with silicone microemulsions. In *Proceedings of the Eighth International Congress on Deterioration and Conservation of Stone: Berlin, 30 September–4 October 1996*, ed. Joseph Riederer, 1215–21. Berlin: Möller Druck und Verlag, 1996. Tested cubes of sandstone impregnated with soluble salts. Solutions form two phases when in contact with salts, one viscous, the other less viscous but weaker.

Costa, D., and J. Delgado Rodrigues. Assessment of color changes due to treatment products in heterochromatic stones. In *The Conservation of Granitic Rocks*, ed.

J. Delgado Rodrigues and D. Costa, 95–101. Lisbon: LNEC, 1996. Ethyl silicate found to have an intermediate darkening effect on granite.

Degradation and Conservation of Granitic Rocks in Monuments: Proceedings of the EC Workshop Held in Santiago de Compostela (Spain) on 28–30 November 1994, ed. M. A. Vicente, J. Delgado Rodrigues, and J. Acevedo. Brussels: European Commission, Directorate-General XII, Science, Research and Development, 1996. Summary of work on the deterioration and conservation of granite. The important articles from this workshop that concern consolidation with alkoxysilanes are abstracted under the names of the individual authors of articles that appeared in *The Conservation of Granitic Rocks* (1996), edited by J. Delgado Rodrigues and D. Costa.

Delgado Rodrigues, J., and D. Costa. Assessment of the efficacy of consolidants in granites. In *The Conservation of Granitic Rocks,* ed. J. Delgado Rodrigues and D. Costa, 63–69. Lisbon: LNEC, 1996. Report on absorption characteristics of consolidants and curing behavior. Evaluated impregnation depth by water vapor permeability and ultrasonic velocity. Found that a second application of ethyl silicate successfully increased amount of product in the stone and also increased mechanical strength.

Delgado Rodrigues, J., and D. Costa. Assessment of the harmfulness of consolidants in granites. In *The Conservation of Granitic Rocks,* ed. J. Delgado Rodrigues and D. Costa, 71–78. Lisbon: LNEC, 1996. Report on the harmfulness of consolidants, measured by water vapor permeability, color changes, mechanical properties, and depth of impregnation. Ethyl silicate was found to reduce open porosity and water vapor permeability; also found to darken the stone.

Delgado Rodrigues, J., and D. Costa. Occurrence and behavior of interfaces in consolidated stones. In *The Conservation of Granitic Rocks,* ed. J. Delgado Rodrigues and D. Costa. Lisbon: LNEC, 1996. Discusses the behavior of consolidated rocks (some with ethyl silicate) with respect to thermal and mechanical stresses (including salt crystallization).

Delgado Rodrigues, J., D. Costa, M. Sa da Costa, and I. Eusebio. Behaviour of consolidated granites under aging tests. In *The Conservation of Granitic Rocks,* ed. J. Delgado Rodrigues and D. Costa, 79–85. Lisbon: LNEC, 1996. Report on testing of consolidated granites with cycles of RH/T, salt crystallization, radiation, long-term immersion. Monitored long-term behavior by FTIR and PA-FTIR. Ethyl silicate was found to be sensitive to the aging process.

Delgado Rodrigues, J., D. Costa, and A. P. Ferreira Pinto. Use of water-absorption characteristics for the study of stone treatments. In *The Conservation of Granitic Rocks,* ed. J. Delgado Rodrigues and D. Costa, 21-27. Lisbon: LNEC, 1996. Report on testing of consolidants on granite. Used microdrop absorption times, water-rock contact angles and water absorption by capillarity to determine spatial distribution and depth of penetration.

Delgado Rodrigues, J., D. Costa, and N. Schiavon. Spatial distribution of consolidants in granite stones. In *The Conservation of Granitic Rocks,* ed. J. Delgado Rodrigues and D. Costa, 55–61. Lisbon: LNEC, 1996. Reports on depth of penetration of consolidants in granite by SEM, ultrasonic velocities, water vapor permeability profiles, water absorption, etc. The ethyl silicate consolidant tested was found to occupy pore spaces; it did not form a film.

Frogner, P., and L. Sjöberg. Dissolution of tetraethylorthosilicate coatings on quartz grains in acid solution. In *Proceedings of the Eighth International Congress on*

Deterioration and Conservation of Stone: Berlin, 30 September–4 October 1996, ed. Joseph Riederer, 1233–41. Berlin: Möller Druck und Verlag, 1996. Studied Wacker OH by Raman spectroscopy and found that polymerization was slower at 50% relative humidity (as opposed to 90%). The dissolution rates of TEOS-derived gel on quartz grains were found to increase in acidic (pH 1-5) solutions containing electrolytes (NH_4Cl, KCl, NaCl).

Garcia Pascua, N., M. I. Sánchez de Rojas, and M. Frias. The important role of color measurement in restoration works: Use of consolidants and water repellents in sandstone. In *Proceedings of the Eighth International Congress on Deterioration and Conservation of Stone: Berlin, 30 September–4 October 1996,* ed. Joseph Riederer, 1351–61. Berlin: Möller Druck und Verlag, 1996. Tested an ethyl silicate and epoxy resins on sandstone. The ethyl silicate was found to slightly darken the sandstone.

Goins, E. S., G. Wheeler, D. Griffiths, and C. A. Price. The effect of sandstone, limestone, marble, and sodium chloride on the polymerization of MTMOS solutions. In *Proceedings of the Eighth International Congress on Deterioration and Conservation of Stone: Berlin, 30 September–4 October 1996,* ed. Joseph Riederer, 1243–54. Berlin: Möller Druck und Verlag, 1996. MTMOS solutions are found to be affected by the presence of limestone, marble and soluble salts. Carbonates slow the hydrolysis reaction leading to a base type polymerization.

Goins, E. S., G. Wheeler, and M. Wypyski. Alkoxysilane film formation on quartz and calcite crystal surfaces. In *Proceedings of the Eighth International Congress on Deterioration and Conservation of Stone: Berlin, 30 September–4 October 1996,* ed. Joseph Riederer, 1255–64. Berlin: Möller Druck und Verlag, 1996. Deposits from alkoxysilane solutions on calcite and quartz crystals were examined by optical microscopy and SEM. Wacker OH formed poor films on both substrates.

Fort Gonzalez, R. Effects of consolidates *[sic]* and water repellents on the colour of the granite rock of the aqueduct of Segovia (Spain). In *Degradation and Conservation of Granitic Rocks in Monuments: Proceedings of the EC Workshop Held in Santiago de Compostela (Spain) on 28–30 November 1994,* ed. M. A. Vicente, J. Delgado Rodrigues, and J. Acevedo, 435–40. Brussels: European Commission, Directorate-General XII, Science, Research and Development, 1996. Indicates that color changes produced with ethyl silicate are not great but tend towards darkening similar to wet stone.

Grissom, C. A. Conservation of Neolithic lime plaster statues from 'Ain Ghazal. In *Archaeological Conservation and Its Consequences: Preprints of the Contributions to the Copenhagen Congress, 26–30 August 1996,* ed. A. Roy and P. Smith, 70–75. London: IIC, 1996. Compares *Conservare* OH and Raccanello/MTMOS mixtures as consolidants for lime plaster.

Hansen, E. F., J. Griswold, L. Harrison, and W. S. Ginell. Desalination of highly deteriorated stone: A preliminary evaluation of preconsolidants. In *11th Triennial Meeting, Edinburgh, Scotland, 1–6 September 1996: Preprints (ICOM Committee for Conservation),* ed. J. Bridgland, 798–804. London: James & James, 1996. Sandstone samples deteriorated by cyclic loading and sodium sulfate were treated with several consolidant systems. The effectiveness of the consolidants to provide adequate wet strength to allow for salt removal by immersion was observed. The alkoxysilanes tested did not provide sufficient strength.

Hilbert, G., and E. Wendler. Influence of different consolidating agents on the water vapour diffusion properties of selected stones. In *Proceedings of the Eighth International Congress on Deterioration and Conservation of Stone: Berlin,*

30 September–4 October 1996, ed. Joseph Riederer, 1345–49. Berlin: Möller Druck
und Verlag, 1996. Sandstones were tested with Funcosil (100, 300, and 510) products. Consolidant performance was discussed vis-à-vis the "Aachen" model. These
consolidants formed gel plates that created secondary pores. The gels did not
decrease vapor transport.

Hristova, J., and V. Todorov. Consolidation effect of Wacker-silicones on the properties of
sandy limestone. In *Proceedings of the Eighth International Congress on
Deterioration and Conservation of Stone: Berlin, 30 September–4 October 1996*,
ed. Joseph Riederer, 1195–1201. Berlin: Möller Druck und Verlag, 1996. Studied
Wacker OH and 290L on sandy limestone by mercury porosimetry, SEM, and ultrasonic velocity and found that both products altered physical and mechanical properties of the stone: the pore volume and fluid permeability decreased while the
modulus of elasticity and strength increased. Wacker OH gave greater increases in
strength.

Kimmel, J. Characterization and consolidation of Pennsylvania blue marble, with a case
study of the second bank of the United States, Philadelphia, PA. Masters of Science
thesis, University of Pennsylvania, 1996. Examined alkoxysilanes for the consolidation of Pennsylvania blue marble.

Koblischek, P. The consolidation of natural stone with a stone strengthener on the basis of
poly-silicic-acid-ethylester. In *Proceedings of the Eighth International Congress on
Deterioration and Conservation of Stone: Berlin, 30 September–4 October 1996*,
ed. Joseph Riederer, 1187–93. Berlin: Möller Druck und Verlag, 1996. Considers
health and safety requirements of consolidants and determines that prehydrolyzed,
solvent-free, ethyl silicate–based stone consolidants are safest.

Lukaszewicz, J. W. The influence of preconsolidation with ethyl silicate on soluble salts
removal. In *Proceedings of the Eighth International Congress on Deterioration and
Conservation of Stone: Berlin, 30 September–4 October 1996*, ed. Joseph Riederer,
1203–8. Berlin: Möller Druck und Verlag, 1996. Limestone and sandstone samples
impregnated with 5, 10, and 15% sodium sulfate solutions were later consolidated
with Funcosil OH, 300 and 510. Objects could be desalinated after consolidation,
but more time was needed versus untreated samples.

Lukaszewicz, J. W. The influence of stone preconsolidation with ethyl silicate on deep consolidation. In *Proceedings of the Eighth International Congress on Deterioration
and Conservation of Stone: Berlin, 30 September–4 October 1996*, ed. Joseph
Riederer, 1209–14. Berlin: Möller Druck und Verlag, 1996. Limestone and
sandstone samples were first treated with Funcosil OH, 300 or 510. After the preconsolidation, the samples were further consolidated with Funcosil H, Anhydrosil
Z, epoxy resin, or ACRYLOID B72. Preconsolidation decreased the absorption rate
of the stones and decreased open porosity by an average of 21.4% in limestone and
18.4% in sandstone. The porous structure was changed by the introduction of polysiloxane gel into the pores.

Martin, W. Stone consolidants: A review. In *A Future for the Past*, ed. J. M. Teutonico,
30–49. London: James & James, 1996. Reviews consolidation methods used at
English Heritage. Emphasizes Brethane and the lime method.

Rager, G., M. Payre, and L. Lefèvre. Mise au point d'une méthode de dessalement pour
des sculptures du XIVe siècle en pierre polychromée. In *Le Dessalement des matériaux poreux*, 241–56. 7th Journées d'Etudes de la SFIIC. Champs-sur-Marne: SFIIC,
1996. Tested a number of products for consolidation of polychrome limestone

sculptures that required desalination. Wacker OH was found to give the most favorable results. Wacker H was found to impede removal of the salts.

Rao, S. M., C. J. Brinker and T. J. Ross. Environmental microscopy in stone conservation. *Scanning* 18(7):508–14. An EDTA functional alkoxysilane coupling agent was examined as a passivant for calcite surfaces. Environmental SEM showed that the coupling agent agglomerated unevenly on the surface and was washed away with water.

Snethlage, R., H. Ling, M. Tao, E. Wendler, L. Sattler, and S. Simon. The sandstone of Dafosi: Investigations of deterioration and conservation methods. *Der Grosse Buddha von Dafosi.* Arbeitshefte des Bayerischen Landesamtes für Denkmalpflege 82:220–39. [German and English with Chinese summary] Discusses the use of Wacker OH on this red clay-bearing sandstone.

Stadlbauer, E., S. Lotzmann, B. Meng, H. Rösch, and E. Wendler. On the effectiveness of stone conservation after 20 years of exposure: Case study of Clemenswerth Castle/NW Germany. In *Proceedings of the Eighth International Congress on Deterioration and Conservation of Stone: Berlin, 30 September–4 October 1996,* ed. Joseph Riederer, 1285–96. Berlin: Möller Druck und Verlag, 1996. Baumberg sandstone was treated with Wacker SL in 1975 and Wacker 190S in 1988. No substantial loss of surface was noted 20 years later. In 1988 a treatment comprising several steps was executed: preconsolidation with Funcosil OH, consolidation with a partial application of Funcosil OH, and a hydrophobic treatment of Wacker 190S. The durability of the treated stone was characterized by drill-resistance measurements and water absorption.

Von Plehwe-Leisen, E., E. Wendler, H. D. Castello Branco, and A. F. Dos Santos. Climatic influences on long-term efficiency of conservation agents for stone—a German-Brazilian outdoor exposure program. In *Proceedings of the Eighth International Congress on Deterioration and Conservation of Stone: Berlin, 30 September–4 October 1996,* ed. Joseph Riederer, 1325–32. Berlin: Möller Druck und Verlag, 1996. In situ testing of different consolidants and water repellents in Germany and Brazil.

Wendler, E., A. E. Charola, and B. Fitzner. Easter Island tuff: Laboratory studies for its consolidation. In *Proceedings of the Eighth International Congress on Deterioration and Conservation of Stone: Berlin, 30 September–4 October 1996,* ed. Joseph Riederer, 1159–70. Berlin: Möller Druck und Verlag, 1996. Tested a modified version of Wacker OH that included a fungicide on the volcanic tuff from Easter Island.

1997

Matero, F., and A. Oliver. A comparative study of alkoxysilanes and acrylics in sequence and in mixture. *Journal of Architectural Conservation* 3(2):22–42. Evaluates the effectiveness of applying ethyl silicate, MTMOS, and ACRYLOID B72 on limestone and compares the results when the alkoxysilanes and the acrylic resin are applied in sequence and as a mixture.

Nagy, K. L., R. Cygan, C. S. Scotto, C. J. Brinker, and C. S. Ashley. Use of coupled passivants and consolidants on calcite mineral surfaces. In *Materials Issues in Art and Archaeology V: Symposium Held 3–5 December 1996, Boston, Massachusetts, USA,* ed. P. B. Vandiver, J. R. Druzik, J. F. Merkel, and J. Stewart, 301–6. Pittsburgh: Materials Research Society, 1997. Explores the use of molecular modeling and laboratory synthesis to develop improved passivating agents for the calcite mineral surface based on trimethoxy dianionic form of silylalkylaminocarboxylate, silyl-alkylphosphonate, and the trisilanol neutral form of aminoethylaminopropylsilane.

Scherer, G., and G. Wheeler. Stress development drying of *Conservare* OH. In *Fourth International Symposium on the Conservation of Monuments in the Mediterranean Basin,* vol. 3, ed. A. Moropoulou, F. Zezza, E. Kollias, and I. Papachristodoulou, 355–62. Athens: Technical Chamber of Greece, 1997. Discusses theory of drying stresses in gels and experiments used to measure drying stresses. Results indicate that *Conservare* OH–derived gels are viscoelastic, with low permeability and pore sizes in the nanometer range.

Wendler, E. New materials and approaches for the conservation of stone. In *Saving Our Architectural Heritage: The Conservation of Historic Stone Structures. Report of the Dahlem Workshop, Berlin, 3–8 March 1996,* ed. N. Baer and R. Snethlage, 182–96. New York: Wiley, 1997. Tests elastified and coupling-agent-modified ethyl silicate consolidants.

1998

Jerome, P. S., N. R. Weiss, A. S. Gilbert, and J. A. Scott. Ethyl silicate as a treatment for marble: Conservation of St. John's Hall, Fordham University. *APT Bulletin* 24(1):19–26. *Conservare* OH gave good test results on a coarse-grained calc-schist.

1999

Bradley, S., Y. Shashoua, and W. Walker. A novel inorganic polymer for the conservation of ceramic objects. In *Twelfth Triennial Meeting, Lyon, 29 August–3 September 1999: Preprints (ICOM Committee for Conservation),* ed. J. Bridgland, 770–76. London: James & James, 1999. Silicon zirconium alkoxides are explored for consolidating ceramics.

Cervantes, J., G. Mendoza-Díaz, D. E. Alvarez-Gasca, and A. Martinez-Richa. Application of ^{29}Si and ^{27}Al magic angle spinning nuclear magnetic resonance to studies of the building materials of historical monuments. *Solid State Nuclear Magnetic Resonance* 13(4):263–69. Examines the changes made to deteriorated stone after treatment with ethyl silicate consolidants by using silicon nuclear magnetic resonance.

Grissom, C. A., A. E. Charola, A. Boulton, and M. F. Mecklenburg. Evaluation over time of an ethyl silicate consolidant applied to ancient lime plaster. *Studies in Conservation* 44(2):113–20. *Conservare* OH has been used for consolidation of ancient lime plaster fragments to enable reassembly of five large statues. The success of treatment was verified by modulus of rupture testing, which demonstrated over 300% increase in strength. When viewed with a scanning electron microscope (SEM), the consolidant closely conformed to coccoliths (fossils) in the plaster, and it was uncracked. The SEM showed no detectable changes in the consolidant over the 11-year period in which samples were examined.

Johansson, U., A. Holmgren, W. Forsling, and R. L. Frost. Adsorption of silane coupling agents onto kaolinite surfaces. *Clay Minerals* 34:239–46. DRIFTS and FT-Raman spectroscopies are used to observe the adsorption of alkoxysilanes (aminopropyl- and glycidoxypropyltrimethoxysilanes) onto clay surfaces.

Simon, S., and A.-M. Lind. Decay of limestone blocks in the block fields of Karnak Temple (Egypt): Non-destructive damage analysis and control of consolidation treatments. In *Twelfth Triennial Meeting, Lyon, 29 August–3 September 1999: Preprints (ICOM Committee for Conservation),* ed. J. Bridgland, 743–49. London: James & James, 1999. Ultrasonic tomography was used to determine the distribution of consolidants in the stone blocks treated with ethyl silicate.

2000

Dell'Agli, G., C. Ferone, G. Mascolo, O. Marino, and A. Vitale. Durability of tufaceous
stones treated with protection and consolidation products. In *Proceedings of the
Ninth International Congress on Deterioration and Conservation of Stone, Venice,
19–24 June 2000,* vol. 2, ed. V. Fassina, 379–85. Amsterdam: Elsevier, 2000.
Reports on the evaluation of tuff consolidated with commercial alkoxysilanes with
regard to physical parameters such as capillarity and porosity. The consolidants
used were IPAGLAZE, ANTIPLUVIOL-S, ESTEL 1100, and SILO 111.

Escalante, M., R. Flatt, G. Scherer, D. Tsiourva, and A. Moropoulou. Particle-modified
consolidants. In *Protection and Conservation of the Cultural Heritage of the
Mediterranean Cities: Proceedings of the Fifth International Symposium on the
Conservation of Monuments in the Mediterranean Basin, Sevilla, Spain, 5–8 April
2000,* ed. E. Galan and F. Zezza, 425–29. Lisses: A. A. Balkema, 2000. Novel consolidants are prepared by adding colloidal oxide particles to silica sols to reduce
shrinkage and cracking of gels.

Escalante, M., J. Valenza, and G. Scherer. Compatible consolidants from particle-modified
gels. In *Proceedings of the Ninth International Congress on Deterioration and
Conservation of Stone, Venice, 19–24 June 2000,* vol. 2, ed. V. Fassina, 459–65.
Amsterdam: Elsevier, 2000. Silicate consolidants containing oxide particles such as
silica and titania produce gels with less shrinkage and fewer cracks.

Espinosa Gaitán, J., E. Ontiveros Ortega, R. Villegas Sánchez, and M. Alcalde Moreno.
Evaluation of treatments for the stone of the Córdoba Door of Carmona (Sevilla,
Spain). In *Protection and Conservation of the Cultural Heritage of the
Mediterranean Cities: Proceedings of the Fifth International Symposium on the
Conservation of Monuments in the Mediterranean Basin, Sevilla, Spain, 5–8 April
2000,* ed. E. Galan and F. Zezza, 431–35. Lisses: A. A. Balkema, 2000. Among the
consolidants tested on this limestone (with clay and quartz) are Tegovakon V and
Wacker OH.

Félix, C., P. Ferrari, and A. Quiesser. Déconsolidation par absorption d'eau de grès traités
avec le silicate d'éthyle: Mesures nondestructives de E, G et V. In *Proceedings of
the Ninth International Congress on Deterioration and Conservation of Stone,
Venice, June 19–24, 2000,* vol. 2, ed. V. Fassina, 287–95. Amsterdam: Elsevier,
2000. Swiss molasse consolidated with ethyl silicate experienced large increases in
Young's modulus, shear modulus, and ultrasonic velocity. These increases are lost in
two wet cycles. The authors suggest that consolidation should be followed with a
water repellent treatment to protect the consolidant.

Fort, R., M. C. López de Azcona, and F. Mingarro. Assessment of protective treatments
based on their chromatic evolution: Limestone and granite in the Royal Palace of
Madrid, Spain. In *Protection and Conservation of the Cultural Heritage of the
Mediterranean Cities: Proceedings of the Fifth International Symposium on the
Conservation of Monuments in the Mediterranean Basin, Sevilla, Spain, 5–8 April
2000,* ed. E. Galan and F. Zezza, 437–41. Lisses: A. A. Balkema, 2000. Soiling of
stone treated with alkoxysilane-based products is more rapid than for untreated stone.

Leisen, H., J. Poncar, and S. Warrack. *German Apsara Conservation Project: Angkor Wat.*
Köln: GACP, 2000. Pamphlet on the German conservation of Angkor Wat, including
consolidation of the sandstone with modified ethyl silicate products.

O'Connor, J. The role of consolidants in the conservation of Sydney sandstone buildings.
In *Proceedings of the Ninth International Congress on Deterioration and
Conservation of Stone, Venice, June 19–24, 2000,* vol. 2, ed. V. Fassina, 413–17.

Amsterdam: Elsevier, 2000. Outlines the methodology for testing the effectiveness of consolidants on Maroubra sandstone in Sydney.

Quaresima, R., G. Scoccia, R. Volpe, and G. Toscani. Behaviour of different treated and untreated stones exposed to salt crystallization test. In *Protection and Conservation of the Cultural Heritage of the Mediterranean Cities: Proceedings of the Fifth International Symposium on the Conservation of Monuments in the Mediterranean Basin, Sevilla, Spain, 5–8 April 2000*, ed. E. Galan and F. Zezza, 455–60. Lisses: A. A. Balkema, 2000. Treatments of ethyl silicate and polymethylsiloxane with and without an ethyltinsiloxane catalyst on limestone, granite, and sandstone prove effective in reducing damage by salt crystallization.

Rohatsch, A., J. Nimmrichter, and I. Chalupar. Physical properties of fine-grained marble before and after conservation. In *Proceedings of the Ninth International Congress on Deterioration and Conservation of Stone, Venice, 19–24 June 2000*, vol. 2, ed. V. Fassina, 453–58. Amsterdam: Elsevier, 2000. Evaluates the change in physical properties of weathered marble treated with several types of consolidants: Wacker 290 + ethyl silicate in ethanol; ACRYLOID B72; Wacker VP1321 + ACRYLOID B72; and RC90. Reports that ethyl silicate and ACRYLOID B72 gave increased compressive strength but *reduced* tensile strength up to 40%.

Rolland, O., P. Floc'h, G. Martinet, and V. Vergès-Belmin. Silica bound mortars for the repairing of outdoor granite sculptures. In *Proceedings of the Ninth International Congress on Deterioration and Conservation of Stone, Venice, 19–24 June 2000*, vol. 2, ed. V. Fassina, 307–15. Amsterdam: Elsevier, 2000. Describes the preparation and properties of a repair mortar composed of crushed granite, glass powder, fumed silica, and ethyl silicate. Wacker OH, Wacker OH + ACRYLOID B72, and Motema 30 were tested and compared for water vapor permeability, capillary absorption, porosity, salt crystallization tests, and adhesion.

Villegas Sánchez, R., J. Espinosa Gaitán, and M. Alcalde Moreno. Study of weathering factors and evaluation of treatments for the stones of Santa Maria de la Encarnación Church, Constantina (Sevilla, Spain). In *Proceedings of the Ninth International Congress on Deterioration and Conservation of Stone, Venice, 19–24 June 2000*, vol. 2, ed. V. Fassina, 697–705. Amsterdam: Elsevier, 2000. Two limestones and three sandstones were treated with an ethyl silicate (Tegovakon V). Due to low depth of penetration into one of the sandstone samples, this treatment is not recommended.

Theoulakis, P., and A. Tzamalis. Effectiveness of surface treatments for sedimentary limestone in Greece. In *Proceedings of the Ninth International Congress on Deterioration and Conservation of Stone, Venice, 19–24 June 2000*, vol. 2, ed. V. Fassina, 493–501. Amsterdam: Elsevier, 2000. Ethyl silicates with and without polydimethylsiloxane and polymethylphenylsiloxane (Tegovakon V, RC70, RC80, and RC90) are evaluated on limestone by freeze-thaw testing and UV exposure.

Thickett, D., N. Lee, and S. M. Bradley. Assessment of the performance of silane treatments applied to Egyptian limestone sculptures displayed in a museum environment. In *Proceedings of the Ninth International Congress on Deterioration and Conservation of Stone, Venice, 19–24 June 2000*, vol. 2, ed. V. Fassina, 503–11. Amsterdam: Elsevier, 2000. Reports on longevity and stability of alkoxysilane treatments on Egyptian limestone sculpture in museums. The treatments were found to have stabilized 85% of the objects. Failures were noted for objects with higher salt concentrations.

Wheeler, G., J. Méndez-Vivar, E. S. Goins, S. A. Fleming, and C. J. Brinker. Evaluation of alkoxysilane coupling agents in the consolidation of limestone. In *Proceedings of the Ninth International Congress on Deterioration and Conservation of Stone, Venice, 19–24 June 2000*, vol. 2, ed. V. Fassina, 541–45. Amsterdam: Elsevier, 2000. Alkoxysilane coupling agents improved the performance of ethyl silicate–based consolidants on limestone in modulus of rupture testing.

2002

Borrelli, E., and M. L. Santarelli. *Silicates in Conservation.* Rome: ICCROM and CISTeC, 2002. (Compact Disc). A PowerPoint presentation of the lecture focusing on the manufacturers and suppliers of silicate-based stone conservation products given at the conference in Torino, *I Silicati nella Conservazione,* 13–15 February 2002.

Martin, B., D. Mason, J. M. Teutonico, and S. Chapman. Stone consolidants: Brethane report of an 18-year review of Brethane-treated sites. In *Stone: Stone Building Materials, Construction and Associated Component Systems. Their Decay and Treatment,* ed. J. Fidler, 3–18. English Heritage Research Transactions 2. London: James & James, 2002. Summarizes the results of an eighteen-year survey of consolidation trials using Brethane at a number of English Heritage sites where the monuments are constructed of limestone and sandstone.

Oliver, A. B. The variable performance of ethyl silicate: Consolidated stone at three national parks. *APT Bulletin* 33(2–3):39–44. Three case studies illustrating that the effectiveness of ethyl silicate depends on the original composition and condition of the stone and on the extent to which the causes of deterioration can be removed.

Ruedrich, J., T. Weiss, and S. Siegesmund. Thermal behaviour of weathered and consolidated marbles. In *Natural Stone, Weathering Phenomena, Conservation Strategies and Case Studies,* ed. S. Siegesmund, T. Weiss, and A. Vollbrecht, 255–72. Geological Society Special Publication No. 205. London: Geological Society, 2002. Ethyl silicate consolidants engender only small stresses in marble when heated.

Index

Note: Page numbers followed by the letters *f* and *t* indicate figures and tables, respectively.

Illustration Credits

About the Author

George Wheeler is Director of Conservation Education and Research at Columbia University and Research Scientist at the Metropolitan Museum of Art. He has published extensively on stone conservation. He did his undergraduate work in art history and physics at Muhlenberg College. He later received an M.A. in art history from Hunter College of the City University of New York; a graduate Certificate in conservation from New York University's Conservation Center; and a Ph.D. in chemistry, also from New York University. He is a Fellow of the American Academy in Rome and the International Institute for Conservation.